Essential Mathematics for Life

BOOK 7

Review of Whole Numbers Through Algebra

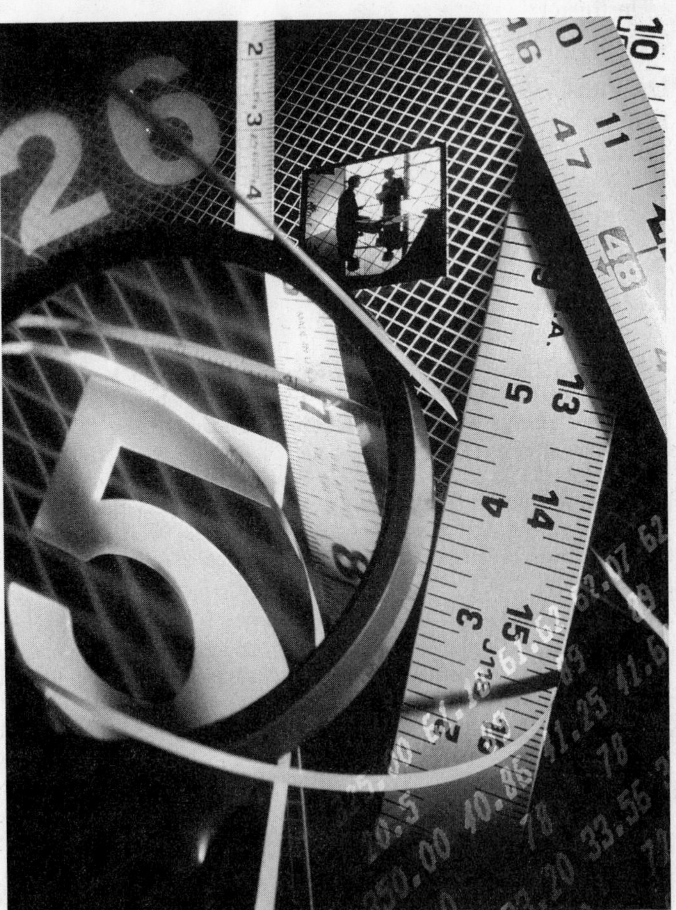

GLENCOE
McGraw-Hill

New York, New York
Columbus, Ohio
Woodland Hills, California
Peoria, Illinois

Authors

Mary S. Charuhas
Associate Dean
College of Lake County
Grayslake, Illinois

Dorothy McMurty
District Director of ABE, GED, ESL
City Colleges of Chicago
Chicago, Illinois

The Mathematics Faculty
American Preparatory Institute
Killeen, Texas

Contributing Writers

Kathryn S. Harr
Mathematics Instructor
Pickerington, Ohio

Priscilla Ware
Educational Consultant and Instructor
Columbus, Ohio

Dr. Pearl Chase
Professional Consultants of Dallas
Cedar Hill, Texas

Contributing Editors and Reviewers

Barbara Warner
Monroe Community College
Rochester, New York

Anita Armfield
York Technical College
Rock Hill, South Carolina

Judy D. Cole
Lafayette Regional Technical Institute
Lafayette, Louisiana

Mary Fincher
New Orleans Job Corps
New Orleans, Louisiana

Cheryl Gunderson
Rusk Community Learning Center
Ladysmith, Wisconsin

Cynthia A. Love
Columbus City Schools
Columbus, Ohio

Joyce Claar
South Westchester BOCES
Valhalla, New York

John Grabowski
St. Joseph Hill Academy
Staten Island, New York

Virginia Victor
Maple Run Youth Center
Cumberland, Maryland

Sandi Braga
College of South Idaho
Twin Falls, Idaho

Maggie Cunningham
Adult Education
Schertz, Texas

Sylvia Gilliard
Naval Consolidated Brig
Charleston, South Carolina

Eva Eaton-Smith
Cecil Community College
Elkton, Maryland

Fabienne West
John C. Calhoun State Community College
Decatur, Alabama

Photo credits: Cover, © Ralph Mercer/Tony Stone Images; 6, Mark Burnett; 9, W. Hill/The Image Works; 12, Glencoe file; 15, L. D. Franga; 18, Glencoe file; 27, Matt Meadows; 29, D. Ogust/The Image Works; 39, Alan Carey/The Image Works; 42, Doug Martin; 45, Bob Daemmrich/Stock Boston; 61, First Image; 64, Gale Zucker/Stock Boston; 66, Aaron Haupt; 79, Doug Martin; 93, Bob Kramer/Stock Boston; 100, Aaron Haupt Photography; 164, Matt Meadows; 186, Morgan Photos; 210, Bob Daemmrich/Stock Boston; 219, Bob Daemmrich/The Image Works; 221, Jules Allen/Vision Photos; 228, Bob Daemmrich/The Image Works; 250, Aaron Haupt Photography; 254, Crown Studios; 274, Tim Courlas; 292, Hickson and Associates.

Printed in the United States of America.

Send all inquiries to:
Glencoe/McGraw-Hill
21600 Oxnard St., Suite 500
Woodland Hills, CA 91367-4906

ISBN: 0-02-802615-2

6 7 8 9 10 11 12 066 02 01 00

CONTENTS

Review

Unit 1 Whole Numbers

Unit 2 Decimals

Unit 3 Fractions

Unit 4 Percents and Proportions

Unit 5 Graphs, Measurements, and Statistics

Unit 6 Introduction to Geometry

Unit 7 Circles

Unit 8 Surface Areas and Volumes

Unit 9 Positive and Negative Numbers

Unit 10 Algebraic Expressions and Equations

Unit 11 Algebra Problem Solving

Unit 12 Exponents, Multiplication, and Factoring

Unit 13 Graphing and Inequalities

UNIT 1

Whole Numbers

Pretest

Write the value of the underlined digit.

1. $612{,}4\underline{8}0$ ____________

2. $\underline{2}40$ ____________

3. $1\underline{5}{,}306$ ____________

4. $1\underline{8}5$ ____________

Write each number in words or numerals.

5. 461

____

6. 2,583

____

7. Two hundred forty-eight

____

8. Nine thousand, fourteen

____

Round to the place value named.

9. 785 to the nearest ten

____

10. 15,642 to the nearest thousand

____

Add.

11. $454 + 27 + 1{,}078 =$ ________

12. $532 + 326 + 1{,}689 =$ ________

13. $3{,}428 + 51{,}235 + 39 =$ ________

14. $83{,}225 + 7{,}453 + 54 =$ ________

Subtract.

15. $\begin{array}{r} 865 \\ -\,360 \\ \hline \end{array}$

16. $\begin{array}{r} 756 \\ -\,638 \\ \hline \end{array}$

17. $\begin{array}{r} 870 \\ -\,424 \\ \hline \end{array}$

18. $\begin{array}{r} 660 \\ -\,378 \\ \hline \end{array}$

19. $\begin{array}{r} 43{,}204 \\ -\,15{,}876 \\ \hline \end{array}$

20. $\begin{array}{r} 90{,}900 \\ -\,78{,}221 \\ \hline \end{array}$

Multiply.

21. $\begin{array}{r} 58 \\ \times\,27 \\ \hline \end{array}$

22. $\begin{array}{r} 75 \\ \times\,37 \\ \hline \end{array}$

23. $\begin{array}{r} 94 \\ \times\,63 \\ \hline \end{array}$

Divide.

24. $5\overline{)2{,}430}$

25. $23\overline{)2{,}254}$

26. $76\overline{)6{,}080}$

Solve the following problems.

27. $2 + 5(3) =$ ________

28. $1 + 8(3) =$ ________

29. $5(4 + 3) =$ ________

30. $22 + 3(8 - 5) =$ ________

Problem Solving

Solve the following problems.

31. Jerry ordered 47 pounds of cheese for a banquet. The order was for 9 pounds of Swiss, 18 pounds of Monterey Jack, and some cheddar. How many pounds of cheddar were ordered? __________

32. A box holds 24 cans of soup. If a store sold 215 boxes, how many cans of soup were sold? __________

33. Cara budgets $287 per month for general office supplies for her business. The budget last year was for $145 per month. How much has the budget increased? __________

34. Carlson College has an intramural soccer league. A total of 440 students signed up for soccer. If each team will have 22 players, how many teams will there be? __________

Place Value

Our number system uses ten digits—0, 1, 2, 3, 4, 5, 6, 7, 8, and 9—to write numbers. The value of a digit depends on its place in the number. This value is called the place value of the digit.

The chart below shows place value. The numbers are separated into groups of threes. A comma is used to separate each group.

Millions				Thousands				Ones		
Hundreds	Tens	Ones		Hundreds	Tens	Ones		Hundreds	Tens	Ones
9	8	2	,	1	7	0	,	5	3	4

comma after the one millions place

comma after the one thousands place

To determine the value of a digit in a number, multiply the digit by the value of its place.

For example: in 982,170,534,

the value of **5** is 5 × 100 or 500,
the value of **1** is 1 × 100,000 or 100,000, and
the value of **8** is 8 × 10,000,000 or 80,000,000.

982,170,534 is read "nine hundred eighty-two million, one hundred seventy thousand, five hundred thirty-four."

MATH HINT

Each group is read by first naming the hundreds, tens, ones, and then naming the group.

Examples

A. Identify the value of **4** in the number 5,746,123.

4 is in the tens place in the thousands group.
The value of **4** is 4 × 10,000 or 40,000.

B. Write **seven thousand, twenty-five** using numerals.

Seven thousand twenty-five has 7 thousands 0 hundreds 2 tens 5 ones.
Using numerals, it is written **7,025.**

 For more information, see Book 1, pages 3–15.

C. Write **5,308** using words.

5,308 has 5 thousands 3 hundreds 0 tens 8 ones.
The number is written **five thousand three hundred eight.**

Practice

Use the chart on page 4 to answer questions 1–4.

1. What is the value of **1** in the number? ____________

2. What is the value of **3** in the number? ____________

3. Write the number using words.

____

4. Write the number using numerals. ____________

Write the value of the underlined digit.

5. 952,<u>7</u>80 ____________

6. <u>8</u>40 ____________

7. 3<u>7</u>,006 ____________

8. 6<u>7</u>5 ____________

9. 3<u>2</u>,060 ____________

10. 81<u>2</u> ____________

Write each number using words or numerals.

11. 294

____

12. 71,983

____

13. 952,780

____

14. 32,060

____

15. 2,475,001

____

16. Three hundred fifty-eight

____

LESSON 2

Rounding Numbers

Sometimes you need to know only "about" how many items you have instead of the exact number. A **rounded number** tells "about" how many. When rounding, you need to know how accurately you want to round and the value the number is closer to.

> **MATH HINT**
>
> Use the "Rule of 5" when you round a number. If the place you are to round is 4 or less, then **round down.** If the place you are to round is 5 or greater, then **round up.**

There were 55,487 people at a baseball game. The sports commentator told her listeners there were about 55,000 people at the game. She rounded the number to the nearest thousand.

Examples

A. Round 2,549 to the nearest hundred.
Ask: Is 2,549 closer to 2,500 or 2,600?
Look at the number to the right of the hundreds place.
It is **4.** $4 < 5$
Round 2,549 to 2,500.

> **MATH HINT**
>
> The symbol $\approx$ means "approximately equal to" and is sometimes used with rounded numbers.

B. Round 798 to the nearest ten.
Ask: Is 798 closer to 790 or 800?
Look at the number to the right of the tens place.
It is **8.** $8 > 5$
Round 798 to 800.

C. Round 10,637 to the nearest thousand.
Ask: Is 10,637 closer to 10,000 or 11,000?
Look at the number to the right of the thousands place.
It is **6.** $6 > 5$
Round 10,637 to 11,000.

 For more information, see Book 1, pages 18–20.

Practice

Round to the place named.

1. 857 to the nearest ten

2. 8,076 to the nearest hundred

3. 56,422 to the nearest thousand

4. 105,986 to the nearest ten

5. 2,196,003 to the nearest ten thousand

6. 97,996 to the nearest hundred

7. 60,402,972 to the nearest ten thousand

8. 56,729,995 to the nearest million

Problem Solving

Solve each of the following problems.

9. Kevin bought a used car for $4,986. When he talked about the cost of the car, he rounded the price to the nearest $1,000. What price did he tell his friend? __________

10. A car has an odometer reading of 52,546 miles. What is the reading to the nearest hundred miles? __________

Shopping

When some people go shopping, they keep track of how much they are spending. Some use a pencil and paper or a calculator to find the exact amount. Others round the prices and estimate.

Louisa needs to buy school supplies. She has $10 to spend. She needs pencils, pens, folders, and notebooks. The price of the items are marked as shown.

Pencils are $0.42 each.
Pens are a box of 12 for $4.89.
Folders are $0.59 each.
Notebooks are 3 for $1.00.

Does Louisa have enough money to buy 4 pencils, a box of 12 pens, 2 folders, and 3 notebooks?

To determine the answer, use rounding.

If pencils are $0.42 each, round the price of each pencil to $0.40. Four pencils would cost about $1.60.
If a box of pens is $4.89, round the cost to $4.90.
If folders are $0.59 each, round the cost to $0.60.
Two folders would cost about $1.20.
Notebooks are 3 for $1.00. When rounded, the cost would still be $1.00.

Now, add the estimated costs of the items.

$$\begin{array}{r} \$1.60 \\ 4.90 \\ 1.20 \\ +1.00 \\ \hline \$8.70 \end{array}$$

The rounded total is $8.70.

Louisa has $10 to spend; therefore, she should have enough money to cover the items plus the tax.

Determine how much money Tyler and Christine need.

1. Tyler wants to buy 3 folders, 4 notebooks, and 5 pencils. About how much money does he need? ____________

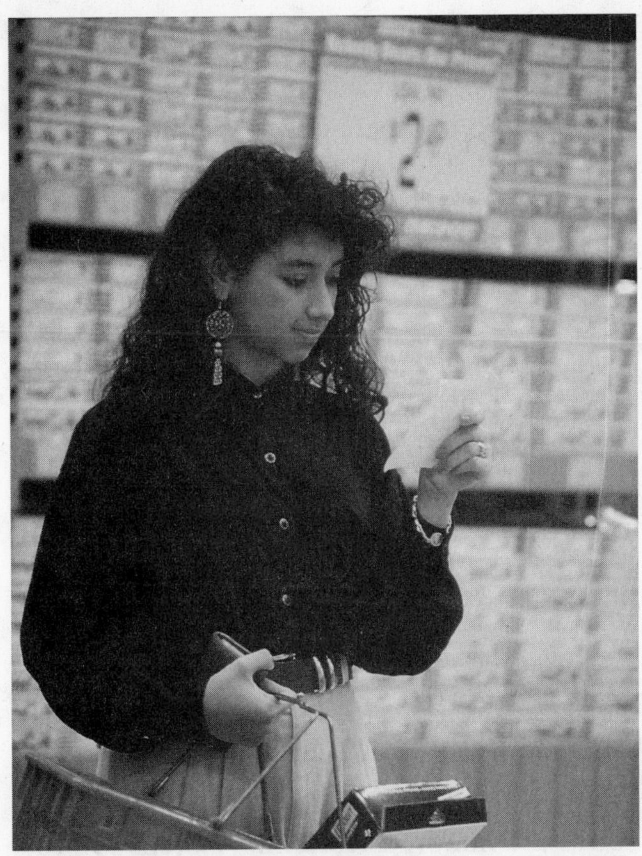

2. Christine needs to buy 6 pencils, a box of pens, 2 notebooks, and 3 folders. About how much money does she need? ____________

Addition

When you add, you put things together. You need to remember to line up the numbers in each place value position and then follow these steps:

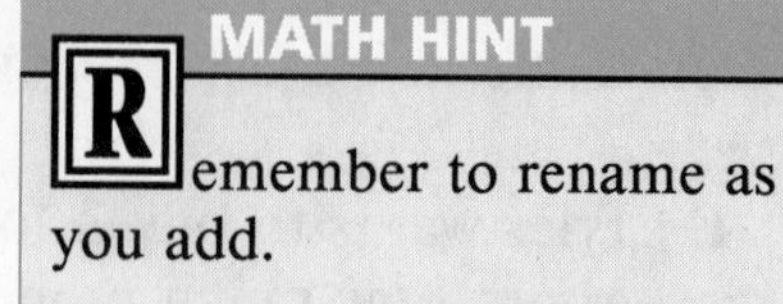

MATH HINT
Remember to rename as you add.

Step 1 Start by adding the numbers in the ones place.
Step 2 Add the numbers in the tens place.
Step 3 Add the numbers in the hundreds place.
Step 4 Continue until there are no numbers left to combine.

Example

A. Add 5,291 and 75,482.
First line up the numbers by place value position.

$$\begin{array}{r} 5,291 \\ +75,482 \\ \hline \end{array}$$

Step 1 Add the numbers in the ones place.
$1 + 2 = 3$

$$\begin{array}{r} 5,291 \\ +75,482 \\ \hline 3 \end{array}$$

Step 2 Add the numbers in the tens place.
$9 + 8 = 17$

$$\begin{array}{r} \scriptstyle 1 \\ 5,291 \\ +75,482 \\ \hline 73 \end{array}$$

Put 7 tens in the tens place in the answer.
Put 1 hundred above the hundreds column. (17 tens is 170)

Step 3 Add the numbers in the hundreds place.
$4 + 2 + 1 = 7$

$$\begin{array}{r} \scriptstyle 1 \\ 5,291 \\ +75,482 \\ \hline 773 \end{array}$$

Step 4 Add the numbers in the thousands place.
$5 + 5 = 10$
Put 0 thousands in the thousands place in the answer.
Put 1 ten thousand above the ten thousands column.

$$\begin{array}{r} \scriptstyle 1\ 1 \\ 5,291 \\ +75,482 \\ \hline 0,773 \end{array}$$

Step 5 Add the numbers in the ten thousands place.
$7 + 1 = 8$

$$\begin{array}{r} \scriptstyle 1\ 1 \\ 5,291 \\ +75,482 \\ \hline 80,773 \end{array}$$

 For more information, see Book 1, pages 32–40.

> **MATH HINT**
>
> Try using graph paper when doing computations by hand. The grid lines help to eliminate alignment errors.

Practice

Add.

1. $\begin{array}{r} 8 \\ +\ 7 \\ \hline \end{array}$

2. $\begin{array}{r} 9 \\ +\ 6 \\ \hline \end{array}$

3. $\begin{array}{r} 15 \\ +\ 34 \\ \hline \end{array}$

4. $\begin{array}{r} 78 \\ +\ 59 \\ \hline \end{array}$

5. $\begin{array}{r} 974 \\ +\ 387 \\ \hline \end{array}$

6. $\begin{array}{r} 7,479 \\ +\ 8,757 \\ \hline \end{array}$

7. 2,006 + 97 + 1,721 = ________

8. 6,566 + 66 + 2,513 = ________

9. 4,283 + 11,215 + 39 = ________

10. 83,225 + 7,453 + 54 = ________

Problem Solving

Solve the following problems.

11. On Monday, Cindy went to the store and spent \$45. On Friday, she spent \$25. How much did she spend this week?

____

12. Sean earns \$350 a week. Last week, he worked the weekend and earned an extra \$60. How much did he earn last week?

____

13. The Jones family took a trip last week. They drove 100 miles on the first day, 300 miles on the second day, and 200 miles on the third day. How many miles did they travel?

____

14. Keith rode his bike 15 miles on Monday, Wednesday, and Friday. He rode 10 miles on Tuesday and Thursday. Over the weekend, he rode 25 miles. How many miles did he ride altogether?

____

15. Renee prepared a report for school. She spent 10 hours researching, 18 hours writing, and 4 hours typing the report. How many hours did she spend preparing this report?

____

Subtraction

You use subtraction to compare two items, find the differences between two items, or reduce the amount of an item. To subtract, line up the numbers in each place value position, putting the larger number on top. Then follow these steps:

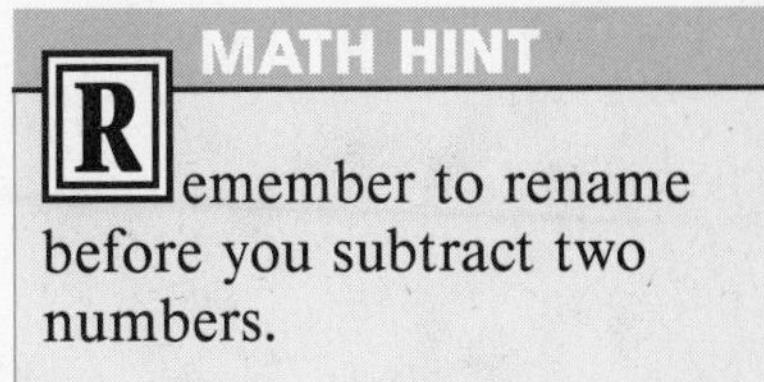

MATH HINT
Remember to rename before you subtract two numbers.

Step 1 Subtract the numbers in the ones place.
Step 2 Subtract the numbers in the tens place.
Step 3 Subtract the numbers in the hundreds place.
Step 4 Continue until there are no numbers left.

Example

A. Find the difference between 50,749 and 39,835. First line up the numbers by place value position. Write the numbers this way:

```
  50,749
- 39,835
--------
```

Step 1
Subtract the ones.

```
  50,749
- 39,835
--------
       4
```

Step 2
Subtract the tens.

```
  50,749
- 39,835
--------
      14
```

Step 3
Rename and then subtract the hundreds.

```
  4 9  17
  5 0 , 7 4 9
- 3 9 , 8 3 5
-------------
        9 1 4
```

Step 4
Subtract the remaining numbers.

```
  4 9  17
  5 0 , 7 4 9
- 3 9 , 8 3 5
-------------
  1 0 , 9 1 4
```

Check your work by adding the answer to the number being subtracted.

10,914 + 39,835 = 50,749

For more information, see Book 1, pages 41–49; 77–90.

Practice

Subtract.

1. $\begin{array}{r} 39 \\ -28 \\ \hline \end{array}$

2. $\begin{array}{r} 26 \\ -12 \\ \hline \end{array}$

3. $\begin{array}{r} 58 \\ -35 \\ \hline \end{array}$

4. $\begin{array}{r} 99 \\ -20 \\ \hline \end{array}$

5. $\begin{array}{r} 83 \\ -59 \\ \hline \end{array}$

6. $\begin{array}{r} 91 \\ -49 \\ \hline \end{array}$

7. $\begin{array}{r} 41 \\ -39 \\ \hline \end{array}$

8. $\begin{array}{r} 82 \\ -66 \\ \hline \end{array}$

9. $\begin{array}{r} 828 \\ -146 \\ \hline \end{array}$

10. $\begin{array}{r} 658 \\ -360 \\ \hline \end{array}$

11. $\begin{array}{r} 567 \\ -386 \\ \hline \end{array}$

12. $\begin{array}{r} 708 \\ -244 \\ \hline \end{array}$

13. $\begin{array}{r} 606 \\ -387 \\ \hline \end{array}$

14. $\begin{array}{r} 32{,}004 \\ -18{,}765 \\ \hline \end{array}$

15. $\begin{array}{r} 99{,}000 \\ -18{,}227 \\ \hline \end{array}$

Problem Solving

Solve the following problems with subtraction.

16. A box of bay leaves originally held 15 leaves. Now, 5 leaves are in the box. How many leaves were used?

____

17. Sara had $7,181 withheld from her paychecks during the year for income taxes. She is due a $453 refund. How much were her income taxes for the year?

____

18. Maggie had 50 dozen eggs to sell. If she had 13 dozen left at the end of the day, how many dozen eggs did she sell?

____

19. This week, Norma spent $165 for groceries. Last week, she spent $195. How much more did she spend for groceries last week?

____

20. Carey deposited $223 into her savings account. She withdrew $75. What is her account balance now?

____

LESSON 5

Multiplication

Think of multiplication as repeated addition. When you multiply, you are actually taking a shortcut to addition. To multiply, line up the place values of each number and follow these steps:

Step 1 Multiply by the digit in the ones place.

Step 2 Then multiply by the digits in each of the remaining places.

Step 3 Add the answers together.

Examples

A. Multiply 83 and 794.

Step 1 794 × 83 = 2,382

$$\begin{array}{r} 794 \\ \times \quad 83 \\ \hline 2,382 \end{array}$$

Step 2 794 × 80 = 63,520

$$\begin{array}{r} 794 \\ \times \quad 83 \\ \hline 2,382 \\ 63,520 \\ \hline \end{array}$$

Step 3 Add the answers.
2,382 + 63,520 = 65,902

$$\begin{array}{r} 794 \\ \times \quad 83 \\ \hline 2,382 \\ 63,520 \\ \hline 65,902 \end{array}$$

To multiply a number by 10, 100, 1,000, and so on, place zeros after the number being multiplied.

B. Multiply 57 by 10.
57 × 10 = 570 Place one zero after 57.

C. Multiply 34 by 1,000.
34 × 1,000 = 34,000 Place three zeros after 34.

MATH HINT

Any number multiplied by one is the same number. Any number multiplied by zero is zero.
345 × 1 = 345
4,567 × 0 = 0

 For more information, see Book 1, pages 121-137.

Practice

Multiply.

1. $\begin{array}{r} 78 \\ \times\ 2 \\ \hline \end{array}$

2. $\begin{array}{r} 65 \\ \times\ 8 \\ \hline \end{array}$

3. $\begin{array}{r} 94 \\ \times\ 4 \\ \hline \end{array}$

4. $\begin{array}{r} 85 \\ \times 76 \\ \hline \end{array}$

5. $\begin{array}{r} 73 \\ \times 27 \\ \hline \end{array}$

6. $\begin{array}{r} 49 \\ \times 36 \\ \hline \end{array}$

7. $689 \times 1 =$ ______

8. $1 \times 250 =$ ______

9. $(1)(736) =$ ______

10. $0 \times 69 =$ ______

11. $591 \times 0 =$ ______

12. $(0)(2{,}136) =$ ______

13. $5 \times 10 =$ ______

14. $6 \times 100 =$ ______

15. $80 \times 1{,}000 =$ ______

16. $\begin{array}{r} 1{,}382 \\ \times\ 906 \\ \hline \end{array}$

17. $\begin{array}{r} 9{,}143 \\ \times\ 502 \\ \hline \end{array}$

18. $\begin{array}{r} 7{,}654 \\ \times\ 308 \\ \hline \end{array}$

Problem Solving

Solve the following problems with multiplication.

19. A store sells boxes of baseballs. Each box contains 30 baseballs. How many baseballs are in 308 boxes?

______

20. Marvin owns 3 coffee shops. Each shop employs 2 clerks and 8 sales-people. How many people does Marvin employ in the 3 shops?

______

21. A bus gets 33 miles per gallon of gasoline. How far can this bus go on 11 gallons of gas?

______

22. A special diet calls for 350 calories for breakfast. How many calories will a dieter consume for breakfast in 7 days?

______

Division

When you divide, you separate a whole into groups. To divide, follow these steps until the problem is done:

Step 1 Divide. Write the digit in the answer above.
Step 2 Multiply.
Step 3 Subtract.
Step 4 Bring down the next digit.

Examples

A. Divide 9,657 by 13.

Step 1 How many 13's in 96? 7
Step 2 Multiply. $7 \times 13 = 91$
Step 3 Subtract. $96 - 91 = 5$
Step 4 Bring down the 5 which is the next digit.

```
     7
13)9,657
   9 1↓
     55
```

Now, repeat the steps.

How many 13's in 55? 4

Multiply. $4 \times 13 = 52$
Subtract. $55 - 52 = 3$

Bring down the 7 which is the next digit.

```
     74
13)9,657
   9 1
     55
     52↓
      37
```

Repeat the steps again.

How many 13's in 37? 2
Multiply. $2 \times 13 = 26$
Subtract. $37 - 26 = 11$

```
     742 R 11
13)9,657
   9 1
     55
     52
      37
      26
      11
```

There is a remainder of 11 in this answer. Place the remainder next to the answer.

B. Divide 4,500 by 5.

$45 \div 5 = 9$ — 4,500 has two zeros.

$4{,}500 \div 5 = 900$ — Add two zeros to 9.

C. Divide 0 by 56.

$0 \div 56 = 0$

> **MATH HINT**
>
> Remember that any number divided into zero is zero. Any number divided by zero is undefined.
>
> $0 \div 34 = 0$
>
> $34 \div 0$ is undefined.

Practice

Divide.

1. $500 \div 1 =$ ______

2. $500 \div 0 =$ ______

3. $500 \div 10 =$ ______

4. $500 \div 5 =$ ______

5. $500 \div 100 =$ ______

6. $560 \div 8 =$ ______

7. $1{,}200 \div 3 =$ ______

8. $8{,}000 \div 40 =$ ______

9. $6{,}400 \div 800 =$ ______

10. $4\overline{)148}$

11. $7\overline{)119}$

12. $8\overline{)4{,}032}$

13. $29\overline{)2{,}262}$

14. $751\overline{)28{,}538}$

15. $205\overline{)1{,}641{,}850}$

Exponents and Roots

A raised smaller number to the right of a number is an **exponent.** An exponent tells how many times a number has been multiplied by itself. The number multiplied by itself is called the **base.**

$3^4 = 3 \times 3 \times 3 \times 3$
or
81 Read "3 to the fourth power is 81."

If there is no exponent, use 1.

$9 = 9^1$ or 9

A number with an exponent of zero equals 1.

$15^0 = 1$

Examples

A. Evaluate 5^3.

$5^3 = 5 \times 5 \times 5$
$5^3 = 125$

To find **the square root of a number** means to find the number that is multiplied by itself to get the number. The symbol used to show this operation is a radical sign $\sqrt{\ }$.

B. Find $\sqrt{36}$.

$\sqrt{36} = 6$ $\quad 6 \times 6 = 36$

This example is a perfect square, which means the square root is a whole number. The square root of a number that is not a perfect square can be found by using a calculator or a table.

For more information, see Book 1, pages 171–172.

Practice

Solve. If necessary, use a calculator.

1. 27^1 __________

2. 6^2 __________

3. 5^4 __________

4. 3^3 __________

5. 10^1 __________

6. 2^5 __________

7. 1^7 __________

8. 8^3 __________

9. $\sqrt{225}$ __________

10. $\sqrt{81}$ __________

11. $\sqrt{1}$ __________

12. $\sqrt{625}$ __________

Order of Operations

A single problem can have many operations. To get the correct answer, the operations must be done in the correct order.

The phrase, "Please Excuse My Dear Aunt Sally," can help you remember the correct order.

P—Parentheses
E—Exponents and roots
MD—Multiplication and Division
AS—Addition and Subtraction

Follow these steps in order to solve problems with several operations.

Step 1 Calculate everything inside the parentheses.

Step 2 Simplify exponents and roots.

Step 3 Calculate multiplication or division operations, moving left to right.

Step 4 Calculate addition or subtraction operations, moving left to right.

Example

A. Solve $(5 + 1) + 2^3(4 - 1)$.

$(5 + 1) + 2^3(4 - 1)$

Step 1 $(6) + 2^3(3)$

Step 2 $(6) + 8(3)$

Step 3 $(6) + 24$

Step 4 30

MATH HINT

Changes in punctuation can greatly change the meaning of a sentence.

Example: "Let's feed the crocodiles, Bob!"
"Let's feed the crocodiles Bob!"

The same is true in mathematical expressions.

Example: $5 - (3 - 2) = 4$
$5 - 3 - 2 = 0$

For more information, see Book 1, pages 173–174.

Practice

Solve the following equations.

1. $3^2(2) =$ __________

2. $5 - 2^2 =$ __________

3. $5^3 - 3^2 =$ __________

4. $9^2 + 6^2 =$ __________

5. $4 + 8(3) =$ __________

6. $5 + 7(3) =$ __________

7. $(2 + 3)(5) =$ __________

8. $(1 + 5)(3) =$ __________

9. $3^2 + 4(8 - 5) =$ __________

10. $5^2 + 3(3 - 2) =$ __________

Problem Solving—Learning to Read the Problem

Read word problems very carefully. Use the following steps to help solve word problems.

Step 1 Read the problem and underline the key words. These words will usually relate to some mathematical reasoning.

Step 2 Make a plan to solve the problem. Ask yourself, Should I add, subtract, multiply, divide, round, or compare? You may have to do more than one of these operations for the same problem.

Step 3 Find the solution. Use your math knowledge to find your answer.

Step 4 Check your answer. Ask yourself, Is the answer reasonable? Did you find what you were asked?

Some key words for word problems are listed below. These words link the problem to an operation. They tell which operation to use.

Addition
sum
total
together
both
increase
more than
plus

Subtraction
difference
subtract
minus
how much more
fewer
reduced by
taken from

Multiplication
product
times
of
twice
triple

Division
quotient
per
average
shared
ratio of
half
divided by

Equals
is
are
was
were
results

For more information, see Book 1, pages 21–22; 48, 91, 116, 136, 165, 180.

Examples

A. In order to raise money for new playground equipment, the residents of a community collected scrap metal to sell to a recycling plant. The amounts of metal collected in pounds for the six weeks were 674, 958, 335, 590, 802, and 629. What was the total amount collected?

Step 1 Determine the total amount collected. The key word is **total.**

Step 2 The key word indicates which operation should occur—addition.

Step 3 Find the solution.

$$674 + 958 + 335 + 590 + 802 + 629 = 3{,}988$$

They collected 3,988 pounds of metal.

Step 4 Check your answer.
Rounding the weights, you get the following problem.

$$670 + 960 + 340 + 590 + 800 + 630 = 3{,}990$$

3,988 and 3,990 are close.

The answer is reasonable.

B. The Ware family drove a van on a three-week summer vacation. They drove 547 miles during the first week. They drove 457 miles during the second week, and they covered 675 miles in the third week. What is the total number of miles driven?

Step 1 Determine the total number of miles driven. The key word is **total.**

Step 2 The key word indicates which operation should occur—addition.

Step 3 Find the solution.

$$547 + 457 + 675 = 1{,}679$$

They drove 1,679 miles in the three weeks.

Step 4 Check your answer.
Rounding the miles driven, you get the following problem.

$$550 + 460 + 680 = 1{,}690$$

1,679 and 1,690 are close.
The answer is reasonable.

Practice

Solve the following problems.

1. A group collected 20 blankets to send to earthquake victims. A box can hold 20 blankets. How many boxes are needed? ____________

2. Sylvia earned $2,518 last summer at camp. She deposited her earnings in a savings account. One year later the account has earned $125 in interest. What is the new balance? ____________

3. A company ordered 47 pounds of colby cheese, 13 pounds of Swiss, 18 pounds of Monterey Jack, 27 pounds of cheddar, and 36 pounds of cream cheese. What was the total amount of cheese ordered? ____________

4. Cookie boxes hold 18 cookies each. An organization sold 1,215 boxes. How many cookies were sold? ____________

5. Janet and Smitty built a float for the annual Fourth of July parade. They bought 78 square yards of material and used only 57 square yards. How many square yards of material were left? ____________

6. A company's budget for office supplies was $1,287 per month. Last year, the office supply budget was $145 less for the year. Find the total amount budgeted for office supplies last year. ____________

7. Petra bought 53 calculators for $24 each. She sold them for $60 each. How much profit did she make if she sold them all? ____________

8. Carlos sold 52 hats at a crafts show. During the first three days, he sold 42 hats at $75 each. On the last day he reduced the price to $60. How much did he earn during the four days? ____________

LIFE SKILL

Expenses

Some power companies offer their customers the option of paying for power by **monthly budget payments.** These payments help customers better budget the costs of the power. The costs for power can vary greatly over a year's time.

Consuela's monthly bills for electric service are listed below.

January	$115	July	$40
February	$98	August	$46
March	$95	September	$55
April	$89	October	$67
May	$57	November	$77
June	$43	December	$92

She paid a total of $874 for electric service. If she pays by budget, her cost would be evenly spread throughout the year. She would pay $73 per month. Look at the payment schedule below.

January	$73	July	$73
February	$73	August	$71
March	$73	September	$73
April	$73	October	$73
May	$73	November	$73
June	$73	December	$73

Notice that the month of August has a different payment. Companies offering budget payments have one month to make adjustments on the budget. The adjustment month is the month in which the actual use and the budgeted use are compared.

Determine the following budget costs.

1. Mark pays $65 a month for gas. His actual use was worth $784 for the year. How much did he pay in the adjustment month? __________

2. Regan pays a monthly budget amount for her electricity. Her actual use for the year was worth $890. Her payment for the adjustment month was $65. How much were her budget payments? __________

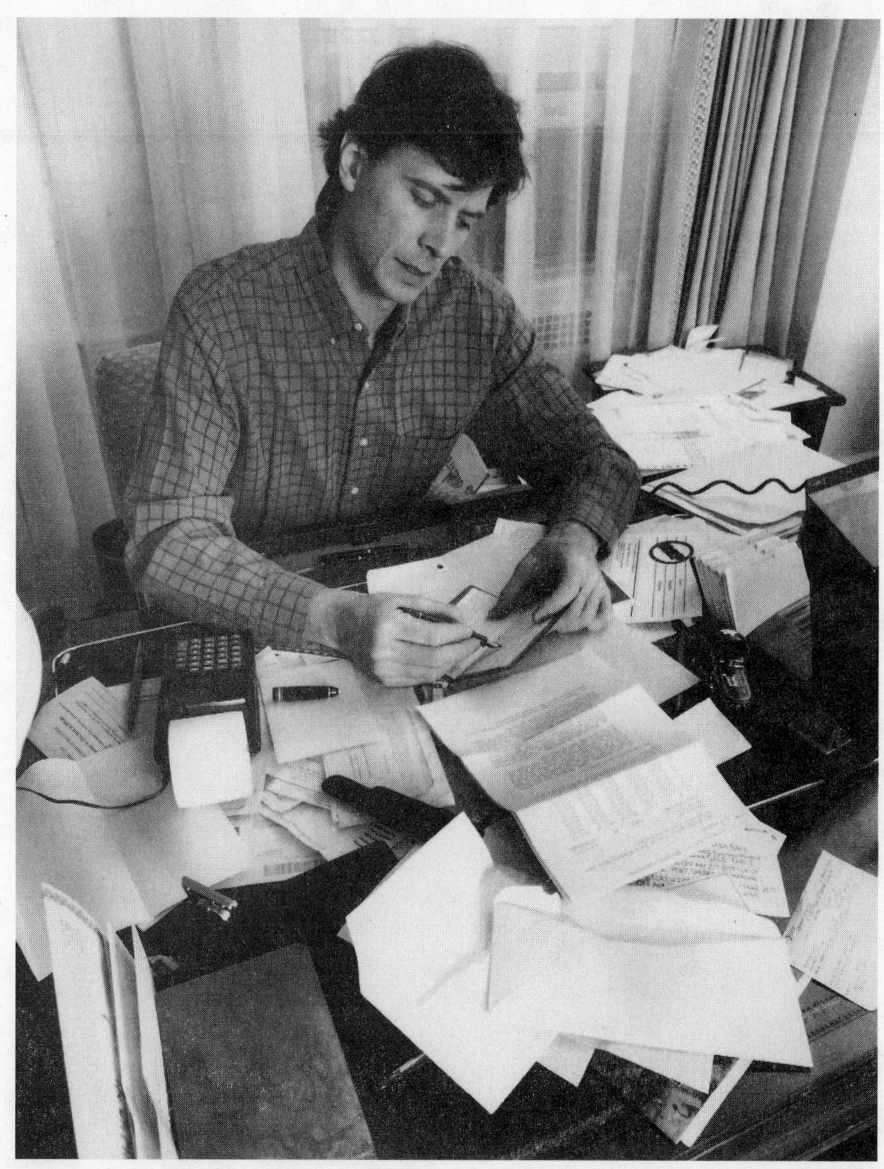

UNIT 1

Posttest

Write the value of the digit underlined.

1. 2,480 ________

2. 520 ________

3. 1,536 ________

4. 518 ________

Write each number in words or numerals.

5. 46 ________

6. 25,183 ________

7. Three hundred seventy-eight ________

8. Nine thousand six hundred four ________

Round to the place value named.

9. 2,784 to the nearest ten ________

10. 35,142 to the nearest thousand ________

11. 2,652,173 to the nearest million ________

Add.

12. 445 + 127 + 8,107 = ________

13. 253 + 632 + 2,168 = ________

Subtract.

14. $\begin{array}{r} 856 \\ -\,630 \\ \hline \end{array}$

15. $\begin{array}{r} 975 \\ -\,836 \\ \hline \end{array}$

16. $\begin{array}{r} 787 \\ -\,449 \\ \hline \end{array}$

17. $\begin{array}{r} 23{,}874 \\ -\,15{,}687 \\ \hline \end{array}$

18. $\begin{array}{r} 89{,}090 \\ -\,78{,}122 \\ \hline \end{array}$

19. 134 − 89 = ________

Multiply.

20. $\begin{array}{r} 38 \\ \times\,37 \\ \hline \end{array}$

21. $\begin{array}{r} 57 \\ \times\,24 \\ \hline \end{array}$

22. $\begin{array}{r} 85 \\ \times\,72 \\ \hline \end{array}$

Divide.

23. $5\overline{)2{,}340}$

24. $67\overline{)5{,}896}$

Problem Solving

Solve the following problems.

25. Maria bought 52 books at a book fair. Forty-eight books cost $135 each. The other books cost $80 each. How much did she spend on books? ________

26. A bus can carry 32 people. How many people can 20 buses carry? ________

27. A typist completed 20 letters in one day. The next day he completed 25 letters. How many letters did he complete in the two days? ________

28. A CD player costs $256. Mike agreed to pay for the CD player by making a down payment of $76 and 6 monthly payments. How much are the monthly payments? ________

UNIT

2

Decimals

Pretest

Write each decimal in words.

1. 0.6

2. 0.22

3. 0.623

4. 7.9

5. 24.6

6. 2.324

Round each number to the place value named.

7. 0.25 (nearest tenth)

8. 0.2136 (nearest thousandth)

9. 9.376 (nearest whole number)

10. 8.607 (nearest hundredth)

Add or subtract.

11. $82.152 + 0.29 =$ ______

12. $93.23 + 52.937 =$ ______

13. $87.31 - 29.8 =$ ______

14. $\begin{array}{r} 56.4 \\ -\,35.958 \\ \hline \end{array}$

Multiply.

15. $0.32 \times 0.06 =$ ______

16. $0.463 \times 6.3 =$ ______

17. $\begin{array}{r} 0.105 \\ \times\,0.32 \\ \hline \end{array}$

18. $\begin{array}{r} 23.46 \\ \times\;\;0.936 \\ \hline \end{array}$

19. $36 \times 20 =$ ______

20. $0.36 \times 0.2 =$ ______

Divide. Round to the nearest hundredth, if necessary.

21. $2.1\overline{)46.2}$

22. $0.84\overline{)21}$

Problem Solving

Solve the following problems.

23. A wire measures 0.3845 inch in diameter. What is the diameter of the wire to the nearest hundredth of an inch?

______

24. Mr. Perry traveled for 3 days. He spent \$402.77, \$398.89, and \$228.24. How much did he spend?

______

LESSON 10

Place Value and Rounding

When you write a price of an item, you use a decimal point to separate the whole dollar from the part of a dollar. If a book costs $4.95, then **4** is the whole dollar and **95** is part of a dollar. The **decimal point** separates the whole and the part.

The first place to the right of the decimal point is the **tenths place**; the second place is the **hundredths**. You can extend numbers beyond two places after the decimal point.

The chart below shows the place values both for numbers greater than 1 and numbers less than 1.

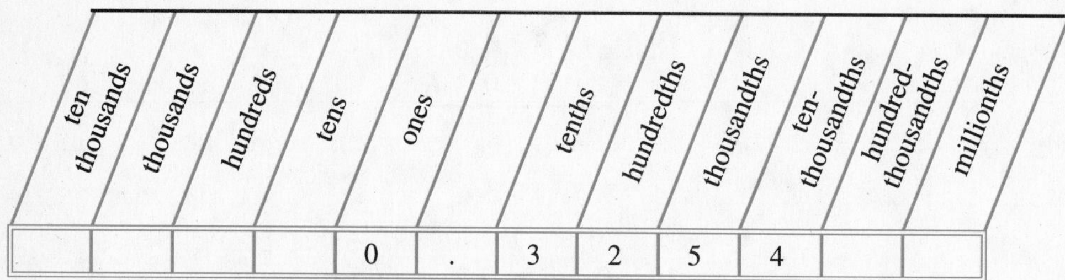

To read the number **0.3254**, first read the numbers after the decimal point as you normally read a number. Then, add the name of the place value in which the last digit is found—**ten-thousandths**. Thus, the number in the chart is read, "three thousand, two hundred fifty-four ten-thousandths."

Examples

A. Read the number 0.455.

0.455 is read "four hundred fifty-five thousandths." When decimals name numbers greater than one, the word **and** is used to locate the decimal point.

B. Read 34.4692.

This number is read "thirty-four and four thousand six hundred ninety-two ten-thousandths."

 For more information, see Book 2, pages 5–12.

When you do not need to know the exact number, you can round decimals. Remember that a rounded number tells "about" how many. When rounding, you need to know how accurately you want to round and the nearest place value of the number.

MATH HINT

Use the "Rule of 5" when you round a number. If the place you are to round is 4 or less, then **round down.** If the place you are to round is 5 or greater, then **round up.**

C. Round $55.47 to the nearest dollar.
Ask: Is $55.47 closer to $55 or $56?
Look at the place to the right of the ones place.
There is a 4. $4 < 5$
Round $55.47 to $55.

D. Round 0.1625 to the nearest hundredth.
Ask: Is 0.1625 closer to 0.16 or 0.17?
Look at the place to the right of the hundredths place.
There is a 2. $2 < 5$
Round 0.1625 to 0.16.

Practice

Write each decimal in words.

1. 0.8

____

2. 0.09

____

3. 0.327

____

4. 9.7

____

5. 68.2543

____

Round each number to the place named.

6. 0.57 (nearest tenth)

____

7. 0.1422 (nearest thousandth)

____

8. 1,509.763 (nearest whole number)

____

9. 8.076 (nearest hundredth)

____

10. 219.6003 (nearest tenth)

____

Addition and Subtraction of Decimals

Adding decimals is like adding whole numbers. To add decimals, follow these steps:

Step 1 Write the numbers in a column, lining up the decimal points. Put a decimal point on the right of any whole number.

Step 2 Add as with whole numbers.

Step 3 Bring the decimal point straight down into the answer.

Examples

A. Add 10.59 and 23.108.

Step 1

```
 10.59
 23.108
 ------
```

Step 2

```
 10.59
 23.108
 ------
 33 698
```

Step 3

```
 10.59
 23.108
 ------
 33.698
```

B. Add 45 and 3.56.

Step 1

```
 45.
  3.56
 -----
```

Step 2

```
 45.
  3.56
 -----
 48 56
```

Step 3

```
 45.
  3.56
 -----
 48.56
```

Subtracting decimals is also like subtracting whole numbers. To subtract decimals, follow these steps:

Step 1 Write the numbers in a column, lining up the decimal points. Put the larger number on the top.

Step 2 If necessary, add zeros as placeholders.

Step 3 Subtract as with whole numbers.

Step 4 Bring the decimal point straight down into the answer.

 For more information, see Book 2, pages 41–54.

Examples

C. Subtract 59.698 from 492.6.

Step 1

$$\begin{array}{r} 492.6 \\ -\ \ 59.698 \\ \hline \end{array}$$

Step 2

$$\begin{array}{r} 492.600 \\ -\ \ 59.698 \\ \hline \end{array}$$

Step 3

$$\begin{array}{r} \overset{8}{4}\overset{11}{\cancel{9}}\cancel{2}.\overset{15}{\cancel{6}}\overset{9}{\cancel{0}}\overset{10}{\cancel{0}} \\ -\ \ 59.698 \\ \hline 432\ \ 902 \end{array}$$

Step 4

$$\begin{array}{r} \overset{8}{4}\overset{11}{\cancel{9}}\cancel{2}.\overset{15}{\cancel{6}}\overset{9}{\cancel{0}}\overset{10}{\cancel{0}} \\ -\ \ 59.698 \\ \hline 432.902 \end{array}$$

D. Subtract 10.59 from 29.4.

Step 1

$$\begin{array}{r} 29.4\ \ \\ -10.59 \\ \hline \end{array}$$

Step 2

$$\begin{array}{r} 29.40 \\ -10.59 \\ \hline \end{array}$$

Step 3

$$\begin{array}{r} 2\overset{8}{\cancel{9}}.\overset{13}{\cancel{4}}\overset{10}{\cancel{0}} \\ -10.59 \\ \hline 18\ \ 81 \end{array}$$

Step 4

$$\begin{array}{r} 2\overset{8}{\cancel{9}}.\overset{13}{\cancel{4}}\overset{10}{\cancel{0}} \\ -10.59 \\ \hline 18.81 \end{array}$$

Practice

Add or subtract.

1. 28.152 + 0.9 = __________

2. 1.666 + 0.03 = __________

3. 90.3 + 29.375 = __________

4. 20.19 + 20.86 + 13.43 = __________

5. 78.13 − 39.1 = __________

6. 0.2 + 0.17 + 0.238 = __________

7. 72.8 − 21.12 = __________

8. 28.152 + 0.9 + 1.666 = __________

9. 66.9 − 34.832 = __________

10. 55.6 − 27.4 = __________

Multiplication With Decimals

Multiplication of decimals is similar to multiplication of whole numbers. When multiplying decimals, follow these steps.

Step 1 Multiply as with whole numbers.

Step 2 Count the number of decimal places in both numbers of the problem and add them together.

Step 3 Count off this total number of decimal places in the answer. Move the decimal point from right to left.

Example

Multiply 0.0534 by 1.3.

Step 1

$$\begin{array}{r} 0.0534 \\ \times \quad 1.3 \\ \hline 1602 \\ 534 \\ \hline 6942 \end{array}$$

Step 2

$$\begin{array}{r} 0.0534 \\ \times \quad 1.3 \\ \hline \end{array}$$

4 places
+ 1 place
5 places
↓
5 places

Step 3

$$\begin{array}{r} 0.0534 \\ 1.3 \\ \hline 1602 \\ 534 \\ \hline 0.06942 \end{array}$$

Practice

Multiply.

1. 0.4 × 0.6 = ________

2. 4.2 × 9.6 = ________

 For more information, see Book 2, pages 59–66; 82–85.

3. $0.03 \times 0.9 =$ __________

4. $6.3 \times 9.2 =$ __________

5. $0.24 \times 0.04 =$ __________

6. $0.634 \times 3.6 =$ __________

7. $0.005 \times 0.12 =$ __________

8. $3.47 \times 0.639 =$ __________

Problem Solving

Solve the following problems.

9. Joe sold 6 spark plugs for a car. Each spark plug costs $1.49. How much did he collect before taxes?

____

10. Ms. Cooke drove an average of 245.85 miles a day for 3.5 days. How many miles did she drive?

____

LESSON 13

Division With Decimals

To divide a decimal by a decimal, follow these steps.

Step 1 If necessary, make the divisor a whole number by moving the decimal point to the right of the last digit.

Step 2 In the dividend, move the decimal point to the right the same number of places.

Step 3 Now place a decimal point directly above in the answer (the quotient).

Step 4 Divide as with whole numbers.

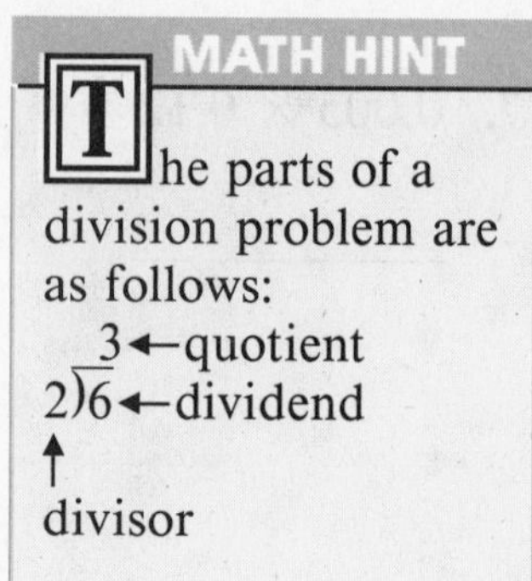

MATH HINT
The parts of a division problem are as follows:
3←quotient
2)6←dividend
↑
divisor

When the division does not come out evenly, divide to one more place than the place to which you are rounding, then apply the rounding rules.

Example

Divide 4.464 by 0.12.

Step 1 0.12)4.464

Step 2 012.)4.464

Step 3 012.)446.4

Step 4

```
      37.2
012.)446.4
     36
      86
      84
       2 4
       2 4
```

 For more information, see Book 2, pages 67-85.

Practice

Divide. If necessary, round to the nearest thousandth.

1. $54 \div 0.03 =$ ______

2. $0.54 \div 3 =$ ______

3. $5.4 \div 300 =$ ______

4. $54 \div 3{,}000 =$ ______

5. $38.7\overline{)132.338}$

6. $5.6\overline{)128.93}$

7. $.015\overline{)0.33}$

8. $.0321\overline{).185}$

9. $2.17\overline{)4.326}$

10. $.7\overline{)5.72}$

LIFE SKILL

Gas Mileage

It is important to keep track of your average number of miles per gallon of gas. Changes in this number can suggest a problem with your car.

To find how many miles your car travels per gallon of gas, you need to keep a record of the number of miles driven and the number of gallons used.

Hint: A full gas tank will give a more accurate reading of miles per gallon.

To determine miles per gallon, follow these steps.

Step 1 Determine the number of miles driven.

Step 2 Determine the number of gallons of gasoline you used to drive those miles.

Step 3 Divide the number of miles driven by the number of gallons used.

Samantha filled her car's gasoline tank 3 days ago with 15 gallons of gasoline. Her car odometer reading was 45,679.6. Today she filled the tank again with 16 gallons of gasoline. The odometer read 46,115.5. Using the steps described on page 42, determine miles per gallon.

Step 1

$$\begin{array}{r} 46{,}115.5 \\ -\ 45{,}679.6 \\ \hline 435.9 \end{array}$$ Miles driven

Step 2 16 gallons
Samantha filled the tank both times. The number of gallons used the second time is how much gasoline she used.

Step 3 435.6 ÷ 16 = 27.225
Samantha's car averaged 27.225 miles per gallon.

Determine the gas mileage for the following problems.

1. Francis drove 525 miles, using 22 gallons of gasoline. What was the average miles per gallon? ________________

2. Lori drove 1,459 miles on her vacation. The car used 45 gallons of gasoline. What was the average miles per gallon on vacation? ________________

Problem Solving—Making a Solution Plan

The steps you have learned to help solve word problems with whole numbers can be used with word problems that deal with decimals. Follow these steps:

Step 1 Read the problem and underline the key words. These words usually relate to some mathematical reasoning computation.

Step 2 Make a plan to solve the problem. Ask yourself, Should I add, subtract, multiply, divide, round, or compare? You may have to do more than one operation for the same problem.

Step 3 Find the solution. Use your math knowledge to find your answer.

Step 4 Check the answer. Ask yourself, Is the answer reasonable? Did you find what you were asked for?

It is important to make a plan to solve the problem. You should ask yourself several questions. What do I want to find? What do I know? How do I find what I need?

Example

Randy bought sporting goods that totaled $84.46. He paid with a $100 bill. How much change should he receive?

Step 1 Determine how much change Randy will receive. The key words are **how much**.

Step 2 This problem asks you to determine the difference between what was spent and what change will be returned from a $100 bill.

Step 3 Find the solution. You will be able to do this by subtracting. You will have to subtract the total charges from $100.

$$\$100 - \$84.46 = \$15.54$$

Randy should receive $15.54 in change.

 For more information, see Book 2, page 12.

Step 4 Check your answer. You can do this by adding the amount of change to the total cost. If the sum is \$100, you know your answer is correct.

$$\$84.46 + \$15.54 = \$100$$

Problem Solving

Solve the following problems.

1. Kent went on a business trip. He spent these amounts: \$202.77, \$198.89, and \$120.24. How much did he spend?

   ____

2. Shay takes a daily count on the gasoline pump to calculate the total number of gallons sold per day. Monday's beginning count was 32,062.5. The ending count was 32,439.3. How many gallons of gasoline were sold on Monday?

   ____

3. Kate had \$1,200.22 in her savings account on July 1. She made two deposits in July: \$770.16 and \$871.93. If she made no withdrawals, how much money does she have in her account?

   ____

4. A city averages 112.5 inches of snowfall. This winter there were 24.32 inches of snow in December, 45.375 inches in January, and 33.8 inches in February. How many inches of snow fell on the city this winter?

   ____

5. Paula increased her babysitter's hourly pay from \$5.65 to \$6.25 per hour. How much more per week does the sitter earn if she works 40 hours in one week?

   ____

UNIT

2

Posttest

Write each decimal in words.

1. 0.3 ______________________

2. 0.72 ______________________

3. 0.363 ______________________

4. 7.29 ______________________

5. 62.46 ______________________

6. 4.925 ______________________

Round each number to the place value named.

7. 0.15 (nearest tenth) __________

8. 0.6238 (nearest thousandth) __________

9. 12.356 (nearest whole number) __________

10. 7.867 (nearest hundredth) __________

Add or subtract.

11. 32.812 + 0.59 = __________

12. 32.59 + 22.58 + 241.43 = __________

13. 78.31 − 59.8 = __________

14. 34.3 − 23.29 = __________

15. $65.4 - 53.558 =$ ________

16. $64.5 - 32.8 =$ ________

Multiply.

17. $0.23 \times 0.08 =$ ________

18. $0.343 \times 5.6 =$ ________

19. $0.315 \times 0.42 =$ ________

20. $43.26 \times 0.136 =$ ________

Divide. Round to the nearest hundredth if necessary.

21. $4.2\overline{)96.6}$

22. $0.68\overline{)38.08}$

23. $0.45\overline{)42.5}$

24. $0.024\overline{)0.33}$

Problem Solving

Solve the following problems.

25. A wooden dowel measures 0.53845 inch in diameter. What is the diameter of the dowel to the nearest hundredth of an inch? ________

26. Margarite made 75 pies for the school bake sale. She sold 0.6 of the pies. How many pies did she sell? ________

UNIT

3

Fractions

Pretest

Reduce each fraction to lowest terms.

1. $\frac{3}{9}$ ________ **2.** $\frac{15}{60}$ ________

3. $\frac{4}{20}$ ________ **4.** $\frac{24}{32}$ ________

Change each improper fraction to a mixed number.

5. $\frac{25}{4}$ ________ **6.** $\frac{31}{5}$ ________

7. $\frac{101}{9}$ ________ **8.** $\frac{93}{7}$ ________

Change each mixed number to an improper fraction.

9. $2\frac{7}{9}$ ________ **10.** $1\frac{7}{10}$ ________

11. $3\frac{1}{3}$ ________ **12.** $2\frac{5}{6}$ ________

Find the least common denominator for each set of numbers. Then rewrite the fractions with the common denominator.

13. $\frac{3}{8}, \frac{1}{2}$ ________ **14.** $\frac{5}{8}, \frac{9}{16}$ ________

Compare the fractions, using >, <, or =.

15. $\frac{5}{9}$ ___ $\frac{4}{9}$

16. $\frac{1}{3}$ ___ $\frac{3}{9}$

17. $\frac{1}{5}$ ___ $\frac{1}{4}$

18. $\frac{5}{1}$ ___ $\frac{11}{2}$

Multiply or divide. Reduce the answers to the lowest terms.

19. $\frac{2}{3} \times \frac{4}{7} =$ ________

20. $\frac{2}{5} \times \frac{6}{8} =$ ________

21. $\frac{5}{9} \div \frac{2}{3} =$ ________

22. $\frac{1}{4} \div \frac{7}{8} =$ ________

23. $3 \times \frac{1}{2} =$ ________

24. $2\frac{1}{9} \times 1\frac{2}{7} =$ ________

25. $1\frac{1}{2} \div \frac{3}{4} =$ ________

26. $5\frac{2}{5} \div 4 =$ ________

Add or subtract. Reduce the answers to the lowest terms.

27. $\frac{2}{8} + \frac{5}{6} =$ ________

28. $\frac{1}{4} + \frac{1}{2} =$ ________

29. $\frac{5}{6} - \frac{2}{3} =$ ________

30. $\frac{7}{18} - \frac{2}{9} =$ ________

31. $3\frac{1}{2} + 2\frac{1}{8} =$ ________

32. $3\frac{1}{5} - 1\frac{2}{9} =$ ________

Problem Solving

Solve the following problems.

33. To make a wall hanging, Karen needs $1\frac{1}{2}$ yards of calico and $\frac{5}{8}$ yards of another fabric. How much fabric does Karen need altogether? ________

34. It takes Ellen $\frac{5}{8}$ of an hour to knit the back of a doll sweater, $\frac{2}{3}$ of an hour to knit the front, and $1\frac{1}{12}$ hours to assemble it. How long does it take Ellen to complete one doll sweater? ________

Fractions and Mixed Numbers

Numbers like 1, 2, 3, and 4 tell how many whole units. Numbers like $\frac{1}{2}$ and $\frac{3}{4}$ are fractions. A **fraction** is part of a whole.

Every fraction has a numerator and a denominator. The **denominator**, or bottom number, tells how many equal parts make a whole. The **numerator**, or top number, tells how many parts of the whole are being used. A short line, or fraction bar, separates the two numbers.

$$\frac{3 \longleftrightarrow}{4 \longleftrightarrow} \quad \frac{\text{numerator}}{\text{denominator}}$$

The fraction $\frac{3}{4}$ tells that the whole was separated into **four parts** and **three** of them were used.

The following examples show the different ways fractions can be written.

1. A **proper fraction** has a numerator that is smaller than its denominator.
   Examples: $\frac{5}{6}$, $\frac{2}{9}$, $\frac{3}{4}$
2. A **mixed number** is a whole number plus a fraction.
   Examples: $2\frac{3}{8}$, $4\frac{5}{7}$, $8\frac{3}{10}$
3. An **improper fraction** has a numerator that is equal to or larger than its denominator. Improper fractions can be rewritten as whole or mixed numbers.
   Examples: $\frac{8}{5}$, $\frac{5}{3}$, $\frac{8}{8}$

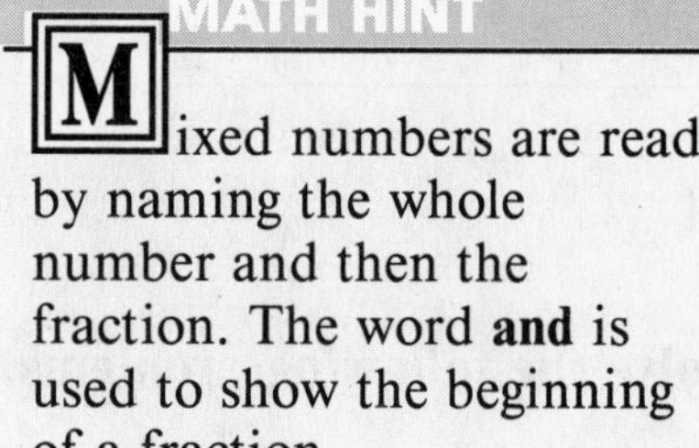

MATH HINT
Mixed numbers are read by naming the whole number and then the fraction. The word **and** is used to show the beginning of a fraction.

Sometimes you need to change a fraction to a different form. For example, you may need to reduce a fraction to lowest terms. To reduce a fraction to lowest terms, follow these steps:

Step 1 Divide both the numerator and the denominator by the same number.

Step 2 Continue to divide until the only number that will divide both the numerator and denominator is **one.**

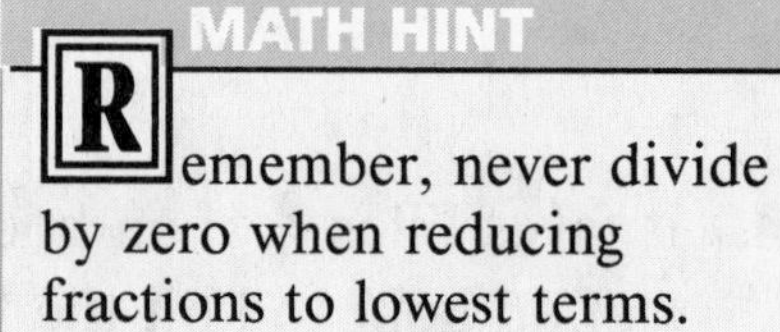

MATH HINT
Remember, never divide by zero when reducing fractions to lowest terms.

 For more information, see Book 2, pages 96–101; 118–122.

Examples

A. Reduce $\frac{4}{8}$ to lowest terms.

Step 1 $\frac{4}{8} = \frac{4 \div 4}{8 \div 4} = \frac{1}{2}$

The answer is $\frac{1}{2}$.

B. Reduce $\frac{16}{36}$ to lowest terms.

Step 1 $\frac{16}{36} = \frac{16 \div 2}{36 \div 2} = \frac{8}{18}$

Step 2 Both 8 and 18 can be divided by 2.

$\frac{8}{18} = \frac{8 \div 2}{18 \div 2} = \frac{4}{9}$

The answer is $\frac{4}{9}$.

To change an improper fraction to a mixed number, follow these steps:

Step 1 Divide the numerator by the denominator.

Step 2 Reduce the answer to lowest terms.

Example

C. Change $\frac{18}{8}$ to a mixed number.

Step 1 Divide 18 by 8.
8 divides into 18 two times with a remainder of 2, so

$$\frac{18}{8} = 8\overline{)18}\quad \begin{array}{r} 2\frac{2}{8} \\ 8\overline{)18} \\ \underline{16} \\ 2 \end{array}$$

Step 2 Reduce the answer to lowest terms whenever possible.

$\frac{18}{8} = 2\frac{2}{8} = 2\frac{1}{4}$

To change a mixed number to an improper fraction, follow these steps.

Step 1 Multiply the whole number by the denominator of the fraction.

Step 2 Add the numerator of the fraction to this product.

Step 3 Write the sum as the numerator of the improper fraction. The denominator remains the same.

Example

D. Change $8\frac{2}{9}$ to an improper fraction.

Step 1 $8 \times 9 = 72$

Step 2 $2 + 72 = 74$

Step 3 74 is the numerator.

Step 4 9 is the denominator.

$$8\frac{2}{9} = \frac{74}{9}$$

Practice

Reduce each fraction to lowest terms.

1. $\frac{6}{9}$ ________ **2.** $\frac{12}{66}$ ________

3. $\frac{5}{20}$ ________ **4.** $\frac{24}{36}$ ________

5. $\frac{20}{24}$ ________ **6.** $\frac{16}{32}$ ________

7. $\frac{6}{30}$ ________ **8.** $\frac{24}{30}$ ________

Change the following improper fractions to mixed numbers.

9. $\frac{22}{4}$ ________ **10.** $\frac{13}{8}$ ________

11. $\frac{27}{8}$ ________ **12.** $\frac{65}{10}$ ________

13. $\frac{28}{6}$ ________ **14.** $\frac{64}{9}$ ________

15. $\frac{25}{4}$ ________ **16.** $\frac{46}{9}$ ________

Change each mixed number to an improper fraction.

17. $2\frac{3}{16}$ ________ **18.** $1\frac{3}{4}$ ________

19. $1\frac{3}{11}$ ________ **20.** $2\frac{1}{3}$ ________

21. $3\frac{2}{3}$ ________ **22.** $1\frac{1}{4}$ ________

23. $3\frac{1}{7}$ ________ **24.** $2\frac{3}{5}$ ________

Common Denominators

When the denominators of fractions are the same, they are called **common denominators**. The fractions $\frac{2}{5}$ and $\frac{3}{5}$ have **5** as the common denominator. However, with fractions, like $\frac{2}{3}$ and $\frac{5}{8}$, you will first have to find a common denominator before working with them.

When you need to find a common denominator for two or more fractions, follow **one** of these steps:

Step 1 Choose a denominator that the other denominator divides into evenly.

or

Step 2 Multiply the denominators together to get a common denominator.

Examples

A. Find the common denominator for $\frac{1}{3}$ and $\frac{4}{9}$.

3 divides evenly into 9.
Use 9 as the common denominator.

B. Find the common denominator for $\frac{1}{3}$ and $\frac{1}{4}$.

$3 \times 4 = 12$
Use 12 as the common denominator.

Practice

Use Step 1 to find a common denominator.

1. $\frac{5}{8}, \frac{1}{2}$ ____________

2. $\frac{5}{8}, \frac{7}{16}$ ____________

3. $\frac{1}{2}, \frac{1}{6}$ ____________

4. $\frac{2}{45}, \frac{7}{15}$ ____________

For more information, see Book 2, page 105.

5. $\frac{1}{14}, \frac{2}{7}$ ____________

6. $\frac{5}{16}, \frac{3}{4}$ ____________

7. $\frac{2}{5}, \frac{4}{15}$ ____________

8. $\frac{2}{3}, \frac{7}{15}$ ____________

9. $\frac{12}{17}, \frac{5}{34}$ ____________

10. $\frac{1}{6}, \frac{5}{36}$ ____________

Use Step 2 to find the common denominator.

11. $\frac{5}{9}, \frac{7}{8}$ ____________

12. $\frac{2}{7}, \frac{7}{9}$ ____________

13. $\frac{1}{13}, \frac{1}{2}$ ____________

14. $\frac{5}{7}, \frac{2}{5}$ ____________

15. $\frac{2}{11}, \frac{1}{3}$ ____________

16. $\frac{5}{7}, \frac{1}{5}$ ____________

17. $\frac{2}{3}, \frac{1}{8}$ ____________

18. $\frac{5}{6}, \frac{3}{7}$ ____________

19. $\frac{2}{15}, \frac{2}{7}$ ____________

20. $\frac{7}{10}, \frac{3}{4}$ ____________

Finding the Least Common Denominator

You should find the least common denominator when adding and subtracting fractions. The **least common denominator** (LCD) is the **smallest** possible common denominator. To find the least common denominator, follow these steps:

Step 1 List multiples of each denominator. A **multiple** of a number is the number multiplied by 1, 2, 3, 4, and so on.

Step 2 Look for the first number in the list that matches a number in the other list. This number will be the **least common denominator**.

Example

Find the least common denominator of $\frac{1}{3}$ and $\frac{3}{8}$.

Step 1 $\frac{1}{3}$ and $\frac{3}{8}$

Multiples of 3	**Multiples of 8**
$3 \times 1 = 3$	$8 \times 1 = 8$
$3 \times 2 = 6$	$8 \times 2 = 16$
$3 \times 3 = 9$	$8 \times 3 = 24$
$3 \times 4 = 12$	
$3 \times 5 = 15$	
$3 \times 6 = 18$	
$3 \times 7 = 21$	
$3 \times 8 = 24$	

Step 2 The first number that matches in each list is 24. 24 is the least common denominator. Notice that 24 is the smallest number that both 3 and 8 can divide into evenly.

Practice

Find the least common denominator for each set of numbers.

1. $\frac{5}{6}, \frac{5}{36}$ ________

2. $\frac{4}{5}, \frac{1}{25}$ ________

3. $\frac{1}{2}, \frac{5}{6}$ ________

For more information, see Book 2, pages 106–107.

4. $\frac{4}{5}, \frac{1}{3}$ ____________

5. $\frac{3}{7}, \frac{1}{42}$ ____________

6. $\frac{7}{16}, \frac{5}{8}$ ____________

7. $\frac{5}{7}, \frac{1}{2}, \frac{9}{14}$ ____________

8. $\frac{5}{8}, \frac{3}{4}, \frac{2}{3}$ ____________

9. $\frac{7}{9}, \frac{3}{4}$ ____________

10. $\frac{3}{4}, \frac{1}{3}, \frac{1}{6}$ ____________

11. $\frac{13}{20}, \frac{1}{5}, \frac{3}{4}$ ____________

12. $\frac{1}{2}, \frac{8}{11}$ ____________

13. $\frac{23}{28}, \frac{5}{7}, \frac{1}{2}$ ____________

14. $\frac{5}{9}, \frac{3}{4}, \frac{1}{3}$ ____________

15. $\frac{1}{4}, \frac{6}{7}$ ____________

16. $\frac{2}{9}, \frac{1}{6}, \frac{3}{18}$ ____________

17. $\frac{2}{5}, \frac{1}{9}, \frac{4}{15}$ ____________

18. $\frac{1}{4}, \frac{1}{3}, \frac{5}{9}$ ____________

19. $\frac{9}{10}, \frac{14}{15}, \frac{1}{6}$ ____________

20. $\frac{2}{3}, \frac{7}{8}, \frac{1}{6}$ ____________

21. $\frac{4}{7}, \frac{3}{5}, \frac{2}{35}$ ____________

Comparing Fractions

To compare fractions, follow these steps:

Step 1 Write all fractions with a common denominator.

Step 2 Compare the numerators.

Example

A. Compare $\frac{4}{5}$ and $\frac{2}{5}$. Use >, <, or =. The denominators are already the same. Compare the numerators.

$4 > 2$

So, $\frac{4}{5} > \frac{2}{5}$.

B. Compare $\frac{1}{3}$ and $\frac{2}{5}$. Use >, <, or =. A common denominator of these fractions is 15.

$\frac{1}{3} = \frac{\quad}{15} \quad \frac{1 \times 5}{3 \times 5} = \frac{5}{15}$

$\frac{2}{5} = \frac{\quad}{15} \quad \frac{2 \times 3}{5 \times 3} = \frac{6}{15}$

Compare the numerators.

$5 < 6$

So, $\frac{5}{15} < \frac{6}{15}$.

$\frac{1}{3} < \frac{2}{5}$

Practice

Compare the following fractions using >, <, or =.

1. $\frac{5}{8}$ ___ $\frac{3}{8}$

2. $\frac{1}{30}$ ___ $\frac{3}{30}$

3. $\frac{1}{3}$ ___ $\frac{1}{4}$

4. $\frac{3}{9}$ ___ $\frac{1}{3}$

5. $\frac{2}{3}$ ___ $\frac{2}{5}$

6. $\frac{5}{8}$ ___ $\frac{7}{8}$

For more information, see Book 2, pages 108–111.

7. $\frac{7}{20}$ ___ $\frac{7}{8}$

8. $\frac{4}{16}$ ___ $\frac{1}{4}$

9. $\frac{6}{9}$ ___ $\frac{12}{18}$

10. $\frac{7}{9}$ ___ $\frac{6}{15}$

11. $\frac{9}{7}$ ___ $\frac{15}{13}$

12. $\frac{14}{5}$ ___ $\frac{17}{8}$

13. $\frac{25}{14}$ ___ $\frac{5}{3}$

14. $\frac{28}{5}$ ___ $\frac{21}{4}$

15. $\frac{18}{5}$ ___ $\frac{15}{4}$

Order the following fractions from the largest to the smallest.

16. $\frac{1}{2}, \frac{1}{6}, \frac{1}{4}$

____

17. $\frac{1}{5}, \frac{3}{5}, \frac{2}{5}$

____

18. $\frac{5}{6}, \frac{4}{5}, \frac{2}{3}, \frac{1}{2}$

____

19. $\frac{6}{7}, \frac{5}{6}, \frac{21}{24}, \frac{5}{7}$

____

20. $\frac{1}{2}, \frac{1}{4}, \frac{1}{5}, \frac{1}{3}$

____

21. $\frac{3}{8}, \frac{1}{4}, \frac{7}{8}, \frac{3}{4}$

____

Order the following fractions from the smallest to the largest.

22. $\frac{5}{11}, \frac{4}{9}, \frac{2}{5}, \frac{3}{7}$

____

23. $\frac{5}{6}, \frac{2}{3}, \frac{3}{4}, \frac{4}{5}$

____

24. $\frac{1}{5}, \frac{2}{3}, \frac{3}{4}, \frac{1}{2}$

____

25. $\frac{2}{4}, \frac{4}{5}, \frac{9}{10}, \frac{1}{8}$

____

Multiplying and Dividing by Fractions

To multiply or divide fractions, it is not necessary to have common denominators.

MATH HINT

When multiplying or dividing fractions, remember to write any mixed numbers as improper fractions.

To multiply a fraction by a fraction, follow these steps.

Step 1 Multiply the numerators together.

Step 2 Multiply the denominators together.

Step 3 Reduce the answer to lowest terms whenever possible.

Examples

A. $\frac{1}{2} \times \frac{3}{4} = ?$

Step 1
Step 2 $\frac{1 \times 3}{2 \times 4} = \frac{3}{8}$

B. $\frac{2}{9} \times \frac{3}{8} = ?$

Step 1
Step 2 $\frac{2 \times 3}{9 \times 8} = \frac{6}{72}$

Step 3 $\frac{\overset{1}{\cancel{6}}}{\underset{12}{\cancel{72}}} = \frac{1}{12}$

MATH HINT

Cancellation makes working with fractions easier. To cancel, divide any numerator and denominator by the same number. This does not change the value of the answer to the problem.

Practice

Multiply. Remember to reduce the answer to lowest terms.

1. $\frac{1}{3} \times \frac{4}{7} =$ ______

2. $\frac{2}{3} \times \frac{6}{8} =$ ______

3. $\frac{7}{21} \times \frac{4}{11} =$ ______

4. $\frac{8}{12} \times \frac{1}{10} =$ ______

5. $\frac{6}{18} \times \frac{2}{12} =$ ______

6. $\frac{1}{6} \times \frac{1}{7} =$ ______

7. $\frac{1}{5} \times \frac{7}{12} =$ ______

8. $\frac{1}{3} \times \frac{1}{10} =$ ______

9. $\frac{5}{10} \times \frac{6}{9} =$ ______

10. $\frac{1}{4} \times \frac{1}{3} =$ ______

11. $\frac{3}{4} \times \frac{7}{12} =$ ______

12. $\frac{2}{9} \times \frac{5}{7} =$ ______

For more information, see Book 2, pages 128–130; 140.

To divide a fraction by a fraction, follow these steps.

Step 1 Invert the divisor. Change the division sign to a multiplication sign.

Step 2 Cancel whenever possible.

Step 3 Follow the rules for multiplication of fractions.

Step 4 Reduce the answer to lowest terms whenever possible.

Example

C. $1\frac{3}{8} \div \frac{5}{8} = ?$

Step 1 $1\frac{3}{8} \times \frac{8}{5}$

Step 2 $\frac{11}{\cancel{8}_1} \times \frac{\cancel{8}^1}{5}$

Step 3 $\frac{11}{1} \times \frac{1}{5} =$

Step 4 $\frac{11}{5} = 2\frac{1}{5}$

Practice

Divide. Remember to reduce the answer to lowest terms.

13. $\frac{5}{12} \div \frac{2}{3} =$ ________

14. $\frac{21}{44} \div \frac{7}{8} =$ ________

15. $\frac{3}{8} \div \frac{10}{5} =$ ________

16. $\frac{3}{8} \div \frac{12}{2} =$ ________

17. $\frac{1}{9} \div \frac{9}{6} =$ ________

18. $\frac{1}{10} \div \frac{10}{6} =$ ________

19. $\frac{6}{7} \div \frac{2}{3} =$ ________

20. $1\frac{7}{8} \div 5 =$ ________

21. $5\frac{1}{4} \div 7 =$ ________

22. $3\frac{3}{4} \div 4\frac{1}{2} =$ ________

23. $2\frac{1}{2} \div 1\frac{1}{2} =$ ________

24. $6\frac{1}{4} \div 2\frac{1}{2} =$ ________

Problem Solving

Solve the following problems.

25. Gracie had $12\frac{1}{2}$ cups of flour to make loaves of bread. If each loaf of bread needs $2\frac{1}{2}$ cups of flour, how many loaves can she make?

______

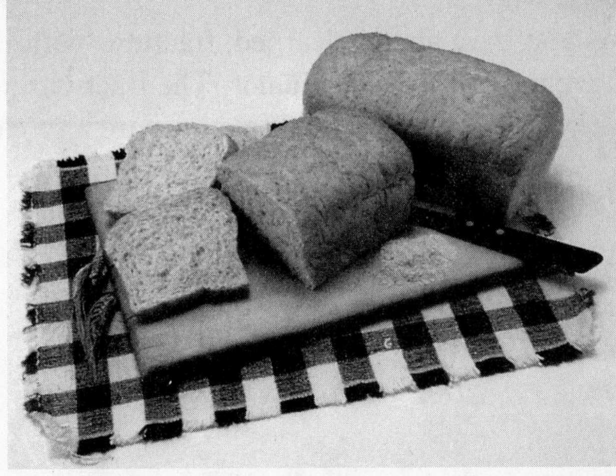

26. Warren mailed 5 packages. If each package weighed $9\frac{1}{2}$ pounds, what was the total weight of the packages?

______

27. A roll recipe calls for $2\frac{1}{4}$ cups of flour. If the recipe is increased by $2\frac{1}{2}$ times, how much flour will be needed?

______

Addition and Subtraction of Fractions

As you have already learned, fractions with the same denominator have a **common denominator**. The fractions $\frac{5}{9}$ and $\frac{4}{9}$ have a common denominator of 9. These fractions are also called **like fractions**.

To add like fractions, follow these steps:

Step 1 Add the numerators.

Step 2 Use the same denominator in the answer.

Step 3 Write the sum in lowest terms.

To subtract like fractions, follow these steps.

Step 1 Subtract the numerators.

Step 2 Use the same denominator in the answer.

Step 3 Write the difference in lowest terms.

Examples

A. $\frac{3}{10} + \frac{1}{10} = \frac{4}{10} = \frac{2}{5}$

B. $\frac{7}{8} + \frac{5}{8} = \frac{12}{8} = 1\frac{4}{8} = 1\frac{1}{2}$

C. $\frac{9}{10} - \frac{1}{10} = \frac{8}{10} = \frac{4}{5}$

D. $\frac{1}{6} - \frac{1}{6} = 0$

Practice

Add or subtract. Write the answers in lowest terms.

1. $\frac{3}{11} + \frac{6}{11}$

2. $\frac{8}{10} + \frac{1}{10}$

3. $\frac{3}{8} + \frac{7}{8}$

4. $\frac{13}{14} + \frac{1}{14}$

5. $\frac{5}{6} - \frac{4}{6}$

6. $\frac{21}{25} - \frac{15}{25}$

7. $\frac{9}{11} - \frac{2}{11}$

8. $\frac{13}{16} - \frac{11}{16}$

 For more information, see Book 2, pages 151–154.

Unlike fractions are fractions with different denominators. Fractions such as $\frac{1}{6}$ and $\frac{3}{10}$ are unlike fractions. Before unlike fractions can be added or subtracted, they must have the same denominator.

To add unlike fractions, follow these steps:

Step 1 Find a common denominator.

Step 2 Write the fractions as equal fractions.

Step 3 Add the numerators.

Step 4 Write the sum in lowest terms.

To subtract unlike fractions, follow these steps.

Step 1 Find a common denominator.

Step 2 Write the fractions as equal fractions.

Step 3 Subtract the numerators.

Step 4 Write the difference in lowest terms.

Examples

A. $\frac{1}{5} + \frac{1}{15}$

$$\frac{1}{5} = \frac{3}{15}$$
$$+\frac{1}{15} = \frac{1}{15}$$
$$\frac{4}{15}$$

Since 15 = 5 × 3, 15 is a common denominator.

MATH HINT

If the smaller denominator divides evenly into the larger denominator, the larger number is a **common denominator**. Otherwise, you can multiply the two denominators to get a common denominator.

B. $2\frac{7}{8} - \frac{1}{3}$

$$2\frac{7}{8} = 2\frac{21}{24}$$
$$-\frac{1}{3} = 2\frac{8}{24}$$
$$2\frac{13}{24}$$

Since 3 does not divide evenly into 8, multiply 3 × 8 to get 24 for a common denominator.

Practice

Add or subtract. Write the answers in lowest terms.

9. $4 + \frac{4}{5}$

10. $2\frac{4}{9} - 1\frac{1}{6}$

11. $\frac{2}{10} - \frac{3}{20}$

12. $4\frac{1}{9} + 2\frac{2}{5}$

13. $5\frac{1}{2} - 4\frac{3}{8}$

14. $2\frac{4}{9} + 2\frac{3}{7}$

15. $3\frac{1}{5} + 1\frac{2}{9}$

16. $5\frac{2}{7} - 3\frac{2}{14}$

LIFE SKILL

Adding Volunteer Hours

Morgan and some of the other parents volunteer reading time to the children who attend Tiny Tots Daycare. Find the total hours each volunteer worked this week.

Volunteers	Day 1	Day 2	Day 3	Day 4	Total Hrs
Morgan	3 hr	$\frac{1}{2}$ hr	1 hr	$\frac{3}{4}$ hr	
Phyllis	$\frac{1}{2}$ hr	2 hr	$\frac{1}{4}$ hr	$\frac{3}{4}$ hr	
Sarah	2 hr	$\frac{3}{4}$ hr	$1\frac{1}{2}$ hr	$\frac{1}{2}$ hr	
Norman	$\frac{1}{4}$ hr	2 hr	$\frac{1}{2}$ hr	$\frac{3}{4}$ hr	
Lucy	1 hr	2 hr	$\frac{3}{4}$ hr	$\frac{1}{4}$ hr	
Gabriel	$\frac{1}{2}$ hr	$\frac{1}{2}$ hr	$\frac{1}{2}$ hr	$\frac{3}{4}$ hr	

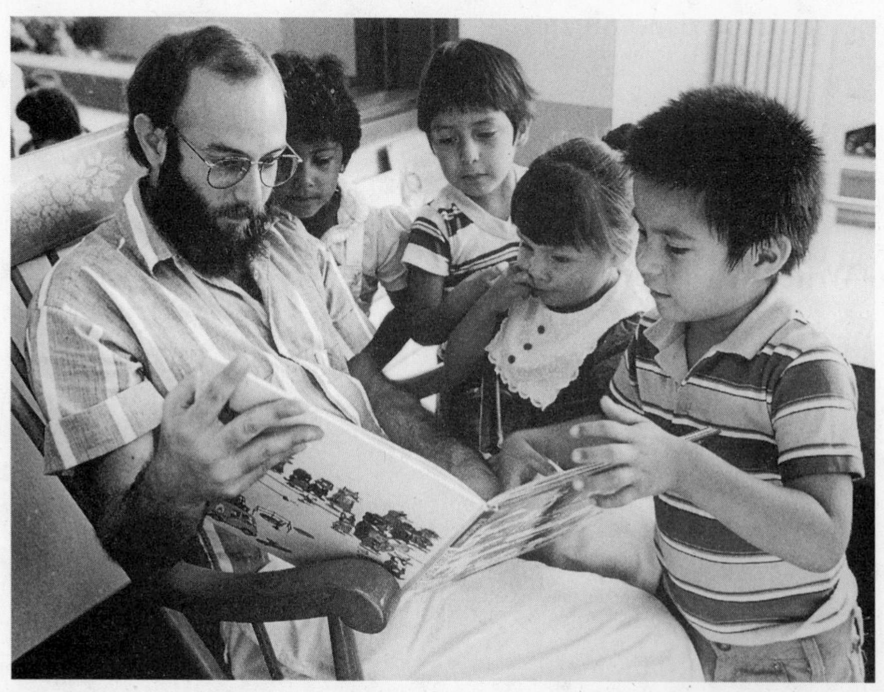

Problem Solving—Answering the Question

The steps you have learned to help solve word problems with whole numbers and decimals can be used with word problems that deal with fractions. Use the following steps:

Step 1 Read the problem and underline the key words. These words will usually relate to some mathematical reasoning.

Step 2 Make a plan to solve the problem. Ask yourself, Should I add, subtract, multiply, divide, round, or compare? You may have to do more than one of these operations for the same problem.

Step 3 Find the solution. Use your math knowledge to find your answer.

Step 4 Check your answer. Ask yourself, Is the answer reasonable? Did you find what you were asked for?

Here is a review of some key words for addition and subtraction.

Addition	**Subtraction**
altogether	decreased by
both	diminished by
increase	difference
sum	how much less
total	how much more
together	remainder

Example

Susie had a piece of material that measured $\frac{3}{4}$ yard long. She cut $\frac{1}{2}$ yard from the material. How much material is left?

Step 1 Determine how much material is left. The key word is **left**.

Step 2 The key word indicates which operation should occur—subtraction.

Step 3 Find the solution.

a. The fractions $\frac{3}{4}$ and $\frac{1}{2}$ are unlike fractions. First, you must find a common denominator for them.

$1 \times 4 = 4 \qquad 1 \times 2 = 2$

$\qquad\qquad\quad\; 2 \times 2 = 4$

The common denominator is **4**.

For more information, see Book 2, page 164.

b. Now, you can subtract, following the steps you learned in Lesson 20.

$$\begin{array}{rr} \frac{3}{4} = & \frac{3}{4} \\ -\frac{1}{2} & \frac{2}{4} \\ \hline & \frac{1}{4} \end{array}$$

Step 4 Check the answer. Does it make sense that $\frac{1}{4}$ would be left? Yes, the answer is reasonable.

Practice

Solve the following problems.

1. Plywood is made by gluing together a backing sheet which is $\frac{3}{8}$ in. thick, an inexpensive core which is $\frac{5}{16}$ in. thick, and a good-grade top sheet which is $\frac{1}{4}$ in. thick. What is the total thickness?

______

2. A student spends $\frac{1}{3}$ of his day sleeping, $\frac{1}{12}$ of his day eating, and $\frac{1}{4}$ of his day in classes. He studies 4 hours each day. How many hours of his day are left?

______

3. The top layer of a quilt uses $1\frac{1}{2}$ yards of blue calico, $2\frac{3}{8}$ yards of red calico, $2\frac{3}{4}$ yards of green calico, and $\frac{5}{8}$ yards of an accent colored fabric. How much fabric is needed for the top layer of the quilt?

______

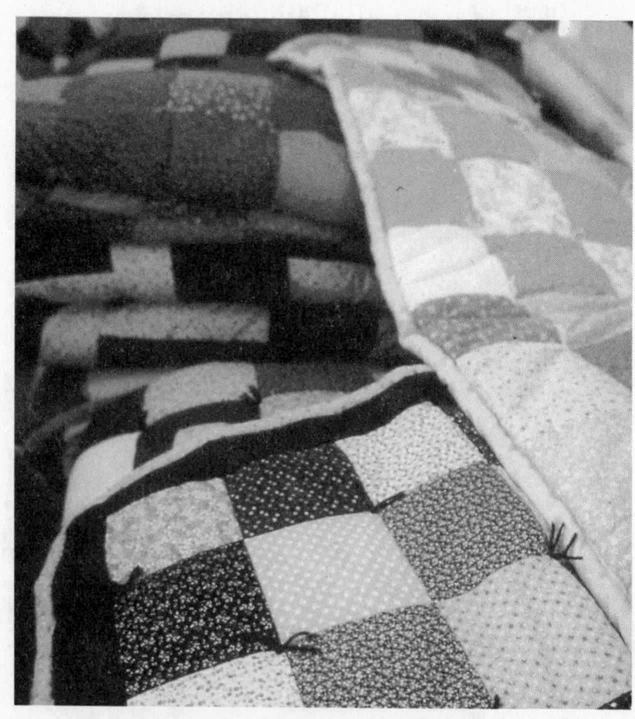

UNIT

3

Posttest

Reduce each fraction to lowest terms.

1. $\frac{2}{4}$ ________

2. $\frac{12}{60}$ ________

3. $\frac{5}{20}$ ________

4. $\frac{16}{32}$ ________

Change each improper fraction to a mixed number.

5. $\frac{25}{3}$ ________

6. $\frac{31}{7}$ ________

7. $\frac{101}{10}$ ________

8. $\frac{93}{8}$ ________

Change each mixed number to an improper fraction.

9. $2\frac{2}{9}$ ________

10. $3\frac{7}{10}$ ________

11. $3\frac{1}{6}$ ________

12. $2\frac{5}{7}$ ________

Find the least common denominator for each set of numbers. Then rewrite the fractions with the common denominator.

13. $\frac{5}{8}, \frac{1}{2}$ ________

14. $\frac{3}{8}, \frac{9}{12}$ ________

Compare the fractions using >, <, or =.

15. $\frac{5}{9}$ ___ $\frac{7}{9}$

16. $\frac{1}{5}$ ___ $\frac{6}{30}$

17. $\frac{1}{6}$ ___ $\frac{1}{5}$

18. $\frac{5}{2}$ ___ $\frac{11}{5}$

Multiply or divide. Reduce the answers to the lowest terms.

19. $\frac{2}{3} \times \frac{3}{7} =$ ______

20. $\frac{2}{5} \times \frac{6}{7} =$ ______

21. $\frac{2}{9} \div \frac{1}{3} =$ ______

22. $\frac{1}{9} \div \frac{5}{8} =$ ______

23. $14 \times \frac{1}{2} =$ ______

24. $2\frac{1}{8} \times 1\frac{2}{5} =$ ______

25. $8\frac{1}{2} \div \frac{3}{4} =$ ______

26. $3\frac{2}{5} \div 4 =$ ______

Add or subtract. Reduce the answers to the lowest terms.

27. $\frac{2}{9} + \frac{5}{6} =$ ______

28. $\frac{3}{4} + \frac{1}{2} =$ ______

29. $\frac{5}{6} - \frac{2}{5} =$ ______

30. $\frac{7}{18} - \frac{2}{6} =$ ______

31. $3\frac{1}{3} + 6\frac{1}{5} =$ ______

32. $7\frac{1}{27} - 3\frac{2}{9} =$ ______

Problem Solving

Solve the following problems.

33. A mail carrier takes $\frac{3}{4}$ of an hour to walk one mile. How many miles does the carrier walk in a 9-hour shift? ______

34. A job takes 20 hours to complete. If Connie has $\frac{1}{4}$ of the job left to complete, how many hours has she already worked? ______

UNIT

4

Percents and Proportions

Pretest

Change to percents.

1. 0.008 = ______ **2.** 0.38 = ______ **3.** 1.15 = ______

4. $\frac{1}{2}$ = ______ **5.** $\frac{1}{6}$ = ______ **6.** $\frac{1}{10}$ = ______

Change to decimals.

7. 15% = ______ **8.** 150% = ______ **9.** 3.5% = ______

Change to fractions.

10. 20% = ______ **11.** 110% = ______ **12.** $12\frac{1}{2}$% = ______

Write each ratio as a fraction and simplify if possible.

13. 12 to 5 ______ **14.** 9 to 21 ______ **15.** 20 to 14 ______

16. 18 to 6 ______ **17.** 14 to 26 ______ **18.** 16 to 20 ______

19. 12 pints to 3 pints ______

20. 6 quarts to 30 pints ______

Solve the following proportion problems.

21. $\frac{5}{7} = \frac{n}{49}$ ______

22. $\frac{n}{42} = \frac{24}{126}$ ______

Read the sentences, then write the fractions as percents.

23. Doug's car payment is $\frac{1}{4}$ of his monthly expenses. ______

24. Judah's rent is $\frac{1}{8}$ of his monthly salary. ______

Problem Solving

Solve the following problems.

25. What is 25% of 200? ______

26. What is 45% of 80? ______

27. 50 is what percent of 25? ______

28. 18 is what percent of 36? ______

29. An item is marked up from $15 to $20. What is the percent of increase in price? ______

30. A small town wants to keep its ratio of 8 parks to 6,000 people. If the town grows to 18,000 people, how many parks will be needed? ______

Percents

A **percent** is a ratio that compares a number to 100. Percent also means *hundredths,* or *per hundred.* The symbol for percent is **%**.

MATH HINT

100% means one whole unit or the entire amount. 0% means none.

Examples

A percent can be written in different ways. The different ways 35% can be written are shown below. The box has been divided into 100 squares, and 35 of 100 squares are shaded.

$\frac{35}{100}$	parts per hundred
	or
$\frac{7}{20}$	reduced fraction
	or
0.35	decimal
	or
35%	percent

$$\frac{35}{100} = \frac{7}{20} = 0.35 = 35\%$$

Practice

Use the figures to answer the following questions.

1. Shade 75 parts of the 100 squares.

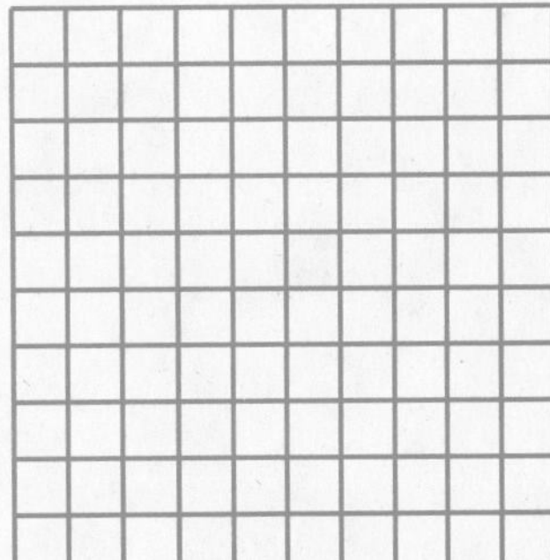

For more information, see Book 3, pages 29–32.

2. Shade 75% of the box.

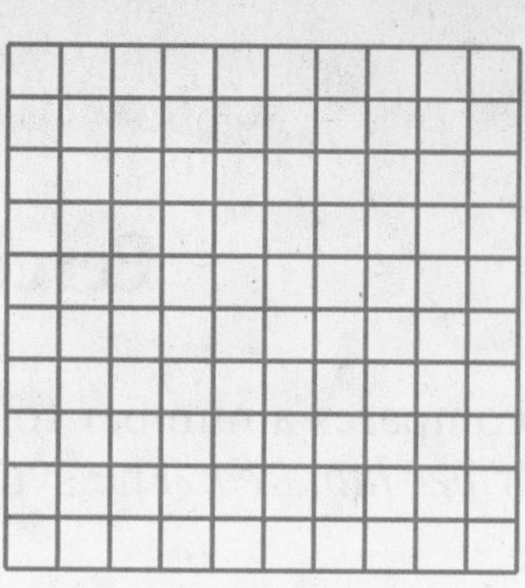

3. Shade 0.75 of the box.

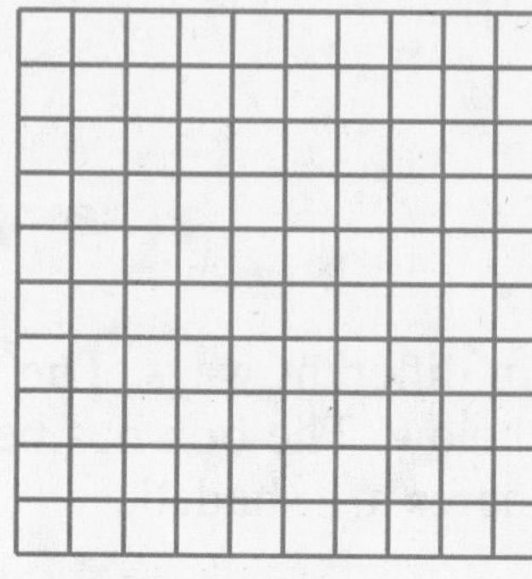

4. Shade $\frac{3}{4}$ of the box.

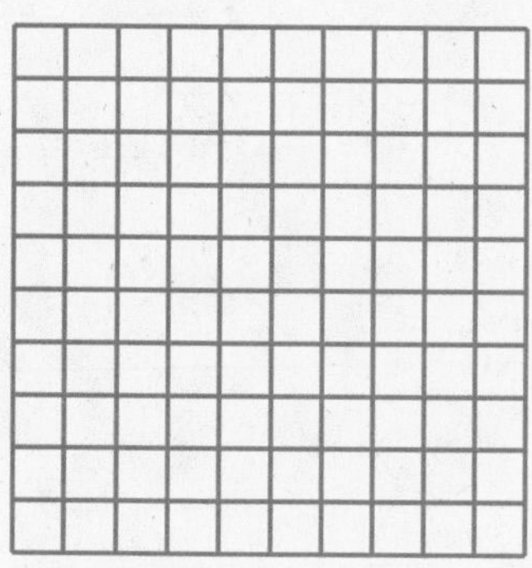

5. Look at the figures above. Compare 75%, 0.75, and $\frac{3}{4}$ by filling in the missing numbers.

$\frac{75}{100} = \frac{3}{\_\_\_\_} = 0.75 = \_\_\_\_\%$

Changing Percents to Decimals, and Decimals to Percents

As you have learned, there are different ways to represent a percent. For example, decimals and percents are two ways to express the same amount.

To change percents to decimals, follow these steps:

Step 1 Remove the % sign.

Step 2 Multiply by 0.01.

MATH HINT

To multiply by 0.01, move the decimal point two places to the left.

75% = 75.0 = 0.75

The result is the same as dividing by 100.

Examples

A. Change 5% to a decimal.

$$\begin{array}{r} 5 \\ \times 0.01 \\ \hline 0.05 \end{array}$$

Step 1 Remove the % sign.

Step 2 Multiply by 0.01.

5% = 0.05

B. Change 300% to a decimal.

$$\begin{array}{r} 300 \\ \times 0.01 \\ \hline 3.00 \end{array}$$

Step 1 Remove the % sign.

Step 2 Multiply by 0.01.

300% = 3.00

C. Change 0.6% to a decimal.

$$\begin{array}{r} 0.6 \\ \times 0.01 \\ \hline 0.006 \end{array}$$

Step 1 Remove the % sign.

Step 2 Multiply by 0.01.

0.6% = 0.006

For more information, see Book 3, pages 54–57.

To change decimals to percents, follow these steps:

Step 1 Multiply the decimal by 100.

Step 2 Add the % sign.

> **MATH HINT**
>
> To multiply by 100, move the decimal point two places to the right.
> 0.75 = 0.75 = 75%

Examples

A. Change 0.34 to a percent.

$$\begin{array}{r} 0.34 \\ \times\ 100 \\ \hline 34.00 \end{array}$$

34%

Step 1 Multiply the decimal times 100.

Step 2 Add the % sign.

> **MATH HINT**
>
> In a whole number, the decimal is located after the ones place or at the right end of the number.

B. Change 2 to a percent.

$$\begin{array}{r} 2.0 \\ \times 100 \\ \hline 200.0 \end{array}$$

200%

Step 1 Multiply the decimal times 100.

Step 2 Add the % sign.

C. Change 0.005 to a percent.

$$\begin{array}{r} 0.005 \\ \times\ 100 \\ \hline 0.500 \end{array}$$

.5%

Step 1 Multiply the decimal times 100.

Step 2 Add the % sign.

Practice

Write the following percents as decimals.

1. 45% ________

2. 198% ________

3. 2% ________

4. 0.7% ________

Write the following decimals as percents.

5. 0.9 ______ 6. 9.0 ______

7. 0.003 ______ 8. 0.65 ______

Problem Solving

Read the sentences below. Then write the percents as decimals, or the decimals as percents.

9. Six percent of Jackie's salary is withheld for dental insurance. ______

10. Gasoline tax has increased by 0.008. ______

11. Maggie saves 15% of her salary. ______

12. The bread had 0.65 walnuts in it. ______

13. A 35% rebate coupon was available for the new soap. ______

14. The dress was made of 0.80 cotton. ______

15. Eighty percent of the visitors to the zoo were children under twelve years of age. ______

Changing Percents, Fractions, and Mixed Numbers

To change percents to fractions, follow these steps:

Step 1 Remove the percent (%) sign. Change to an improper fraction, if necessary.

Step 2 Multiply by $\frac{1}{100}$.

Step 3 Cancel and reduce, if necessary.

Examples

A. Change 5% to a fraction.

Step 1 $5\% \rightarrow 5 \rightarrow \frac{5}{1}$

Step 2 $\frac{5}{1} \times \frac{1}{100}$

Step 3 $\frac{\overset{1}{\cancel{5}}}{1} \times \frac{1}{\underset{20}{\cancel{100}}} = \frac{1}{20}$

B. Change 850% to a fraction.

Step 1 $850\% \rightarrow 850 \rightarrow \frac{850}{1}$

Step 2 $\frac{850}{1} \times \frac{1}{100}$

Step 3 $\frac{\overset{17}{\cancel{850}}}{1} \times \frac{1}{\underset{2}{\cancel{100}}} =$

$\frac{17}{2} = 8\frac{1}{2}$

C. Change $\frac{1}{4}\%$ to a fraction.

Step 1 $\frac{1}{4}\% \rightarrow \frac{1}{4}$

Step 2 $\frac{1}{4} \times \frac{1}{100} = \frac{1}{400}$

D. Change $33\frac{1}{3}\%$ to a fraction.

Step 1 $33\frac{1}{3}\% \rightarrow 33\frac{1}{3} \rightarrow \frac{100}{3}$

Step 2 $\frac{100}{3} \times \frac{1}{100}$

Step 3 $\frac{\overset{1}{\cancel{100}}}{3} \times \frac{1}{\underset{1}{\cancel{100}}} = \frac{1}{3}$

Practice

Write the following percents as fractions. Reduce to lowest terms.

1. 60% __________

2. 25% __________

3. $7\frac{1}{3}\%$ __________

4. 50% __________

 For more information, see Book 3, pages 58–61.

5. $15\frac{1}{2}\%$ ____________

6. 180% ____________

7. $66\frac{2}{3}\%$ ____________

8. $22\frac{1}{5}\%$ ____________

9. 12% ____________

10. 95% ____________

Read the sentences. Then write the percents as fractions. Reduce to lowest terms.

11. Skirts are on sale for 35% off.

____

12. Forty percent of the class passed the test.

____

13. Bologna is 15% fat free.

____

14. Nine percent of the class are women.

____

15. Twelve percent of the workers have two children.

____

16. There is a 200% markup on the cost of electronics equipment.

____

To change fractions to percents, follow these steps:

Step 1 Multiply the fraction by $\frac{100}{1}$.

Step 2 Cancel and reduce. Write any remainders as fractions.

Step 3 Add the % sign.

Examples

A. Change $\frac{1}{4}$ to a percent.

Step 1 $\frac{1}{4} \times \frac{100}{1}$

Step 2 $\frac{1}{\cancel{4}_1} \times \frac{\cancel{100}^{25}}{1} = \frac{25}{1}$

Step 3 25%

B. Change $5\frac{1}{4}$ to a percent.

Step 1 $5\frac{1}{4} \times \frac{100}{1}$

Step 2 $\frac{21}{\cancel{4}_1} \times \frac{\cancel{100}^{25}}{1} = \frac{525}{1}$

Step 3 525%

C. Change $\frac{1}{500}$ to a percent.

Step 1 $\frac{1}{500} \times \frac{100}{1}$

Step 2 $\frac{1}{\cancelto{5}{500}} \times \frac{\cancelto{1}{100}}{1} = \frac{1}{5}$

Step 3 $\frac{1}{5}\%$

D. Change $\frac{1}{3}$ to a percent.

Step 1 $\frac{1}{3} \times \frac{100}{1}$

Step 2 $\frac{1}{3} \times \frac{100}{1} = \frac{100}{3} = 33\frac{1}{3}$

Step 3 $33\frac{1}{3}\%$

Practice

Write the following fractions as percents. Reduce to lowest terms.

17. $\frac{2}{5}$ __________

18. $\frac{29}{10}$ __________

19. $\frac{1}{35}$ __________

20. $\frac{5}{7}$ __________

21. $\frac{4}{5}$ __________

22. $\frac{5}{25}$ __________

23. $\frac{2}{3}$ __________

24. $\frac{1}{8}$ __________

25. $\frac{1}{2}$ __________

26. $\frac{9}{10}$ __________

Read the sentences. Then write the fractions as percents.

27. The material in the skirt was $\frac{2}{5}$ silk.

____

28. The interest rate is $\frac{6}{25}$ of the amount owed.

____

29. The birth rate increased by $\frac{1}{6}$.

____

30. Food prices increased by $\frac{1}{9}$.

____

31. Pay raises were decreased this year by $\frac{1}{7}$.

____

32. The material in the skirt was $\frac{2}{3}$ silk.

____

LIFE SKILL

Discounts

Duncan and Sherry work at the Ace Appliance Shop, which is having its annual sale. All appliances will be sold at discount. Before placing sales tags on items, they worked out a discount list that shows the fraction discounts for each item. To double-check the sales prices they decided to use percent discounts.

Help them change the fraction discounts to the equivalent percent discounts. The first one is done for you.

Appliance	Fraction Discount	Percent Discount
Television	$\frac{1}{3}$	$33\frac{1}{3}\%$
VCR	$\frac{1}{8}$	
Washer	$\frac{1}{4}$	
Dryer	$\frac{2}{5}$	
Toaster	$\frac{1}{2}$	
Camera	$\frac{1}{5}$	
Computer	$\frac{1}{6}$	
Mixer	$\frac{2}{3}$	
Electric knife	$\frac{3}{4}$	

Recognizing Percent Problems

Percent problems are made up of four elements: the **whole**, the **part**, the **percent**, and **100**.

The **whole** is the number value after the word **of**.
The **part** is listed before or just after the word **is**.
The **percent** always has a percent sign **(%)**.
All percents are based on **100**.

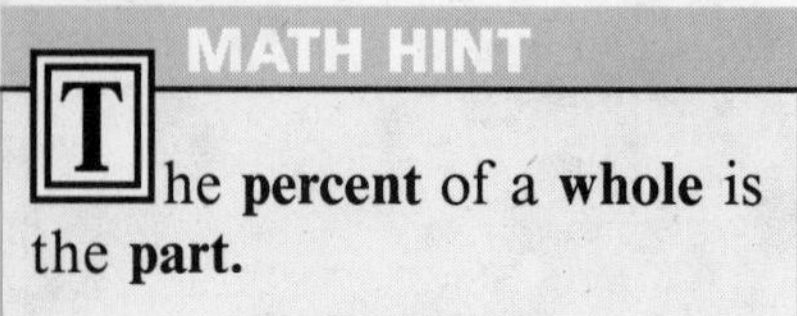
MATH HINT

The **percent** of a **whole** is the **part**.

Example

50% of 300 is 150.

percent of **whole** = **part**

The percent is 50; it has a **percent sign**.

The whole is 300; it is the number after the word **of**.

The part is 150; it is after the word **is**.

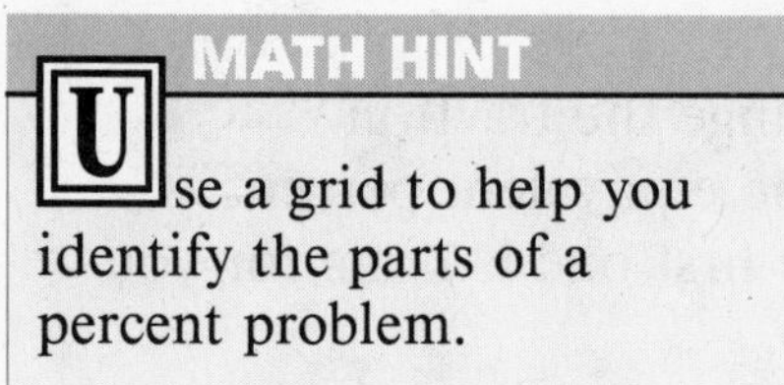
MATH HINT

Use a grid to help you identify the parts of a percent problem.

Practice

For each of the following percent problems, place the part, whole, percent, and 100 on the grid.

1. 75% of 400 is 300.

part	percent
whole	100

2. 1.5% of 100 is 1.5.

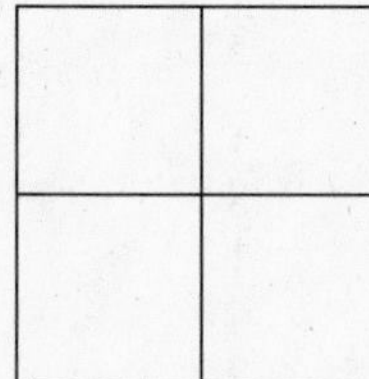

3. 6 is $33\frac{1}{3}$% of 18.

4. 133 is 20% of 665.

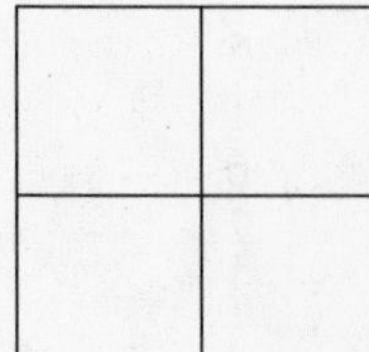

 For more information, see Book 3, pages 70–72; 77–78.

Solving for the Part, Percent, or Whole

When solving for the part, percent, or whole, you can use one of two methods—(1) the **decimal** method or (2) the **fraction** method.

MATH HINT

The decimal method works well with a calculator or long division. On the other hand, the fraction method uses cancellation.

When using the **decimal** method to solve for the **part**, follow these steps:

Step 1 Identify the parts. Place the known information on a grid.

Step 2 Multiply the shaded diagonals.

Step 3 Divide the answer by the number that is left using long division.

When using the **fraction** method to solve for the **part**, follow these steps:

Step 1 Identify the parts. Place the known information on a grid.

Step 2 Multiply the shaded diagonals. (This number becomes the numerator.)

Step 3 Divide the answer by the number that is left. (This number becomes the denominator.)

Example

A. What is 30% of 770?

Step 1

? part	30 percent
770 whole	100 100

Step 2 $770 \times 30 = 23{,}100$

Step 3

```
        231
100)23,100
    20 0
     3 10
     3 00
       100
       100
```

The answer is 231.

30% of 770 is 231.

B. What is 50% of 250?

Step 1

? part	50 percent
250 whole	100 100

Step 2 50×250

Step 3 $\frac{(\overset{1}{\cancel{50}} \times \overset{125}{\cancel{250}})}{\underset{\underset{1}{\cancel{2}}}{\cancel{100}}} = \frac{125}{1} = 125$

The answer is 125.

50% of 250 is 125.

For more information, see Book 3, pages 79–102.

Practice

Place the known information on the grid. Solve for the part, using the decimal method.

1. What is 99% of 300?

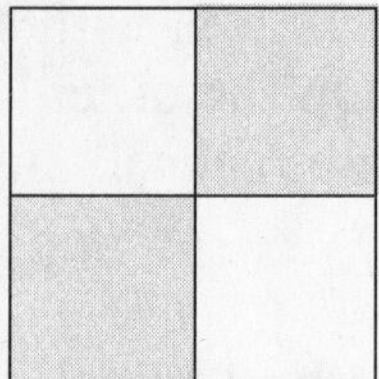

______

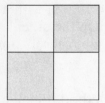

2. What is 10% of 5,500?

______

Place the known information on the grid. Solve for the part, using the fraction method.

3. What is 70% of 70?

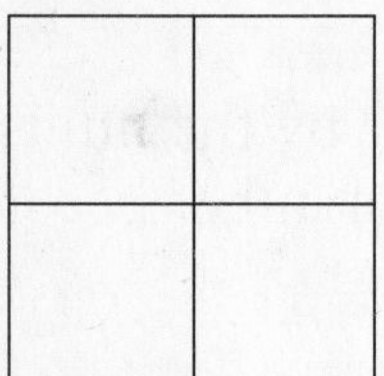

______

4. What is 20% of 80?

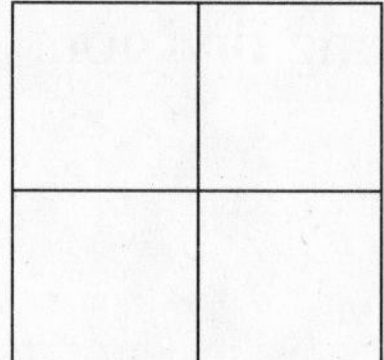

______

When using the **decimal** method to solve for the **percent**, follow these steps:

Step 1 Identify the parts. Place the known information on a grid.

Step 2 Multiply the shaded diagonals.

Step 3 Divide the answer by the number that is left.

Step 4 Add the % sign.

When using the **fraction** method to solve for the **percent**, follow these steps:

Step 1 Identify the parts. Place the known information on a grid.

Step 2 Multiply the diagonals. (This number becomes the numerator.)

Step 3 Divide the answer by the number that is left. (This number becomes the denominator.)

Step 4 Add the % sign.

Examples

C. 13 is what percent of 25?

Step 1

13	?
part	percent
25	100
whole	100

Step 2 $13 \times 100 = 1{,}300$

Step 3

$$\begin{array}{r} 52 \\ 25\overline{)1{,}300} \\ \underline{1\ 25} \\ 50 \\ \underline{50} \end{array}$$

Step 4 52%
13 is 52% of 25.

D. 96 is what percent of 640?

Step 1

96	?
part	percent
640	100
whole	100

Step 2 96×100

Step 3 $\frac{96 \times 100}{640} = \frac{9600}{640} = 15$

Step 4 15%
96 is 15% of 640.

Practice

Place the known information on the grid. Solve for the percent, using the decimal method.

5. 6 is what percent of 15?

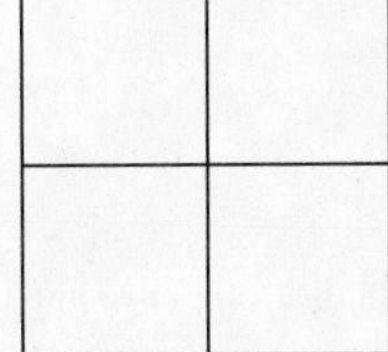

______

6. What percent of 2,880 is 720?

______

Place the known information on the grid. Solve for the percent using the fraction method.

7. What percent of 500 is 10?

______

8. 48 is what percent of 12?

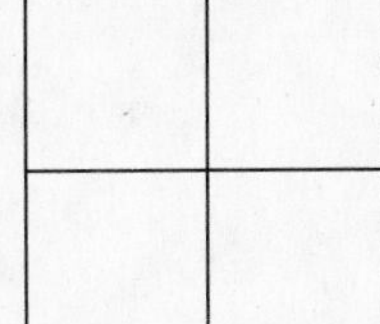

______

When using the **decimal** method to solve for the **whole,** follow these steps:

Step 1 Identify the parts. Place the known information on a grid.

Step 2 Multiply the shaded diagonals.

Step 3 Divide the answer by the number that is left.

When using the **fraction** method to solve for the **whole,** follow these steps:

Step 1 Identify the parts. Place the known information on a grid.

Step 2 Multiply the shaded diagonals. (This number becomes the numerator.)

Step 3 Divide the answer by the number that is left. (This number becomes the denominator.)

Examples

65% of what is 130?

Step 1

130 part	65 percent
? whole	100 100

Step 2 $130 \times 100 = 13{,}000$

Step 3

$$\begin{array}{r} 200 \\ 65\overline{)13{,}000} \\ \underline{13\,0} \\ 00 \\ \underline{00} \end{array}$$

The answer is 200.
65% of 200 is 130.

175% of what is 770?

Step 1

770 part	175 percent
? whole	100 100

Step 2 770×100

Step 3 $\dfrac{\overset{110}{\cancel{770}} \times \overset{4}{\cancel{100}}}{\underset{1}{\cancel{\underset{\cancel{7}}{175}}}} = 440$

The answer is 440.
175% of 440 is 770.

Practice

Place the known information on the grid. Solve for the whole, using the decimal method.

9. 20% of what is 133?

______

10. 38 is 100% of what?

______

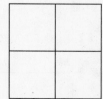

Place the known information on the grid. Solve for the whole, using the fraction method.

11. 75% of what is 18? ______

12. 30% of what is 21? ______

13. 23% of what is 92? ______

14. 88 is 11% of what? ______

15. 40% of what is 20? ______

16. 60% of what is 30? ______

Adding and Subtracting Percents

A **whole unit** can be grouped into many parts. No matter how many groups are created, the whole unit is always 100%. Sometimes a figure can help you better understand a problem.

Example

A. Thirty-five percent of the people in a new community live in condos, 30% live in two-story houses, and the rest live in ranch homes. What percent live in ranch homes?

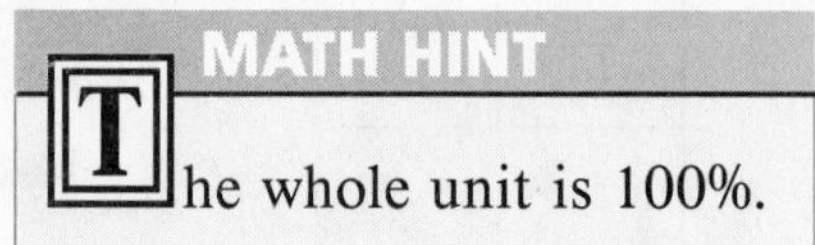
MATH HINT
The whole unit is 100%.

Draw a figure to help you see the math problem.
Divide the figure into three parts, each part representing the three different groups. Label one part 35% for the number of people who live in condos, the second part 30% for those who live in two-story houses, and an unknown percent for those living in ranch homes. Label the whole figure 100%.

65%		?%	100%
35% Condos	30% Two-story Houses	Ranch Homes	

The figure helps show that to **find the percent of people living in ranch homes**, you must:

1. Add the percent of those living in two-story houses to the percent of those living in condos.

35%	Condos
30%	Two-story Houses
65%	

2. Subtract the total from 100%.

$$\begin{array}{r} 100\% \\ -\ 65\% \\ \hline 35\% \end{array}$$

35% of the people in this community live in ranch homes.

 For more information, see Book 3, pages 131–133.

Practice

Solve the following problems. Use the figures provided to help you see the problem.

1. Employees work one of three shifts at the hospital. Thirty-three percent of the employees work the day shift and 45% work the night shift. What percent work the swing shift? ________

2. Twenty-two percent of the people who attended the state fair lived within ten miles of the fairgrounds. Thirty-three percent of the people who attended came from other parts of the state. What percent of people were from out of state? ________

3. In the new apartment building, 15% of the apartments have one bedroom, 44% have two bedrooms, and the rest will have three bedrooms. What percent will have three bedrooms? ________

4. Twenty-five percent of all sales in the toy department were made between 4:00 p.m. and 7:00 p.m. Thirty percent were made from 6:00 p.m. to 8:30 a.m. What percent were made during the rest of the day? ________

5. Thirty percent of Joe's time at work was spent packing boxes. Twenty percent of his time was spent loading the delivery truck. The rest of the time was spent delivering. What percent of his time was spent delivering? ________

Multiplying and Dividing Percents

Some percent problems may ask you to multiply or divide. Drawing and labeling a figure may help you to understand and solve a problem.

Example

Charlie and his three brothers paid for a cruise for their parents' fiftieth wedding anniversary. Charlie paid 40% of the cost of the cruise; each of his three brothers paid an equal share of the rest. If the cruise cost $1,500, how much did each of the three brothers pay?

To solve this problem, follow these steps:

Step 1 Draw a figure.

Step 2 Mark off 40% for Charlie's share.

Step 3 Then divide the remainder by 3 for the three brothers' share. Label the remaining part of the figure Brother 1, Brother 2, and Brother 3.

40%	20%	20%	20% 100%
Charlie's share	Brother 1	Brother 2	Brother 3

Since the entire amount is 100%, the three brothers' share is 60%.

$$\begin{array}{r} 100\% \\ -\ 40 \\ \hline 60\% \end{array}$$

The question asks what each of Charlie's three brothers paid for the cruise. $60\% \div 3 = 20\%$

Step 4 Now the question can be simplified to the phrase: **20% of $1,500 is what?**

Step 5 Use a grid to solve the problem.

? part	20 percent
1,500 whole	100 100

Multiply the diagonals, and divide by the number that is left.

$$\frac{20 \times \overset{15}{\cancel{1{,}500}}}{\underset{1}{\cancel{100}}} = 300$$

Each of Charlie's three brothers paid $300.

 For more information, see Book 3, pages 137–138.

Practice

Solve the following problems by first labeling the diagram. Then use the grid method to solve the problem.

1. Ray Vaca took a 150-mile bike tour that lasted two days. How many miles did he travel on the second day if he covered 30% of the trip on the first day?

2. When Lydia moved, she rented a van to move her belongings. Her belongings weighed 4,000 pounds. The van could hold only 43% of her things. If the rest of her belongings were divided equally between two cars, how much weight was in each car?

3. Max paid 35% of his parents' anniversary party. His five sisters paid an equal share of the rest. If the party cost $600, what amount did each sister pay?

Percent of Increase or Decrease

In percent problems in which you find the percent of increase or decrease, you are comparing the difference between two numbers. The question usually asks, **"What is the percent of increase or decrease . . .?"**

> **MATH HINT**
> The **amount of the difference** is the **part** in the problem. The **original amount** is the **whole.**

Use diagrams to help solve percent of increase or decrease problems. In a diagram, the figure representing the original amount and the figure representing 100% are always the same size.

Example

This year's attendance at the Home and Garden Show was 300,000. Last year, 250,000 people attended. What is the percent of increase for attendance?

Step 1 Determine the original amount. In this problem, it is 250,000.

Step 2 Determine the new amount. It is 300,000.

Step 3 Subtract to find the difference between the new amount and the original amount.

Step 4 The difference becomes the part. The difference between 300,000 and 250,000 is 50,000.

Step 5 Solve for the percent by using the grid method.

The whole is last year's attendance.
The part of the whole is the difference between last year's attendance and this year's attendance.

$$\text{Part} = 300{,}000 - 250{,}000 = 50{,}000$$

 For more information, see Book 3, pages 145–147.

50,000 part	? percent
250,000 whole	100 100

Multiply the diagonals and divide by the number that is left.

$$\frac{\overset{20}{\cancel{50{,}000}} \times \overset{1}{\cancel{100}}}{\underset{1}{\cancel{250{,}000}}_{\cancel{2{,}500}}} = \frac{20}{1} = 20\%$$

Practice

Solve the following problems. Use a grid to help you.

1. The price of meat changed from $4.50 a pound to $5.75 a pound. By what percent did the price increase?

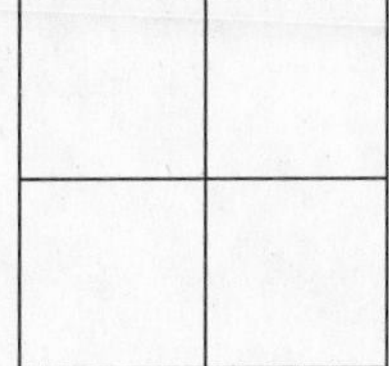

______

2. VCRs decreased in price from $450 to $350. What is the percent of decrease?

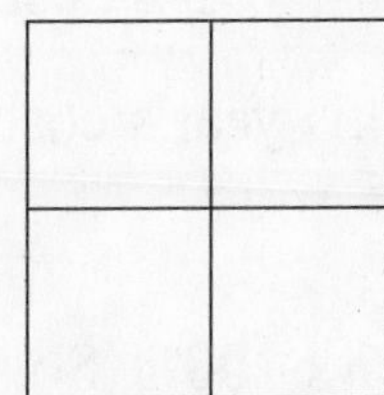

______

3. The sale price of a CD player was $330. The original cost was $420. What was the percent of decrease? Round your answer to the nearest one percent.

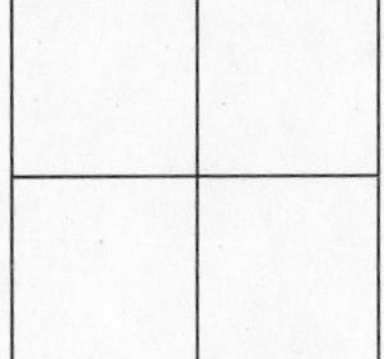

______

4. Jenna's new winter coat cost $75. This year, it cost $95. What is the percent of increase? Round your answer to the nearest one percent.

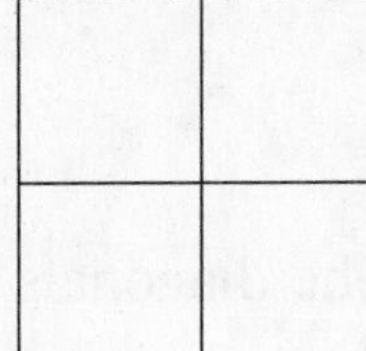

______

LESSON 30

Amount of Increase or Decrease

Knowing the percent of change can also help you to determine the new amount of increase or decrease.

Example

Janice's cost for her children's day care this year was \$1,500. This is 20% more than last year. What was last year's day care cost?

	100%
Last year's cost	20%
This year's cost	\$1,500
	120%

The original amount is always 100%. So, 100% appears at the end of the original cost.

The increase was 20%. Add 20% to 100%.
100% + 20% = 120%
120% would appear under \$1,500.

This year's cost is 120% of what?
\$1,500 is 120% of what?

Use the grid to solve.

1,500	120
part	percent
?	100
whole	100

Multiply the diagonals and divide by the number that is left.

$$\frac{1{,}500 \times \overset{10}{\cancel{100}}}{\underset{12}{\cancel{120}}} = \frac{15{,}000}{12} = 1{,}250$$

The cost of last year's day care was \$1,250.

 For more information, see Book 3, pages 150–151.

Practice

Solve the following problems.

1. Juan makes $5.62 an hour. This is 20% more than he earned per hour last month. What was last month's hourly rate?

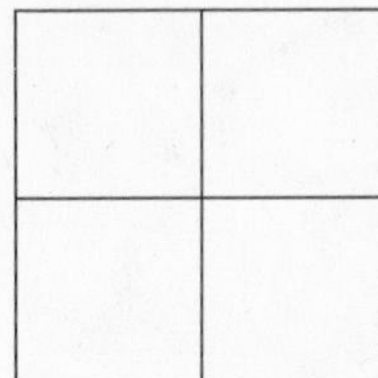

____

2. Clare bought a coat for $250. This was 20% less than the original price. What was the original price?

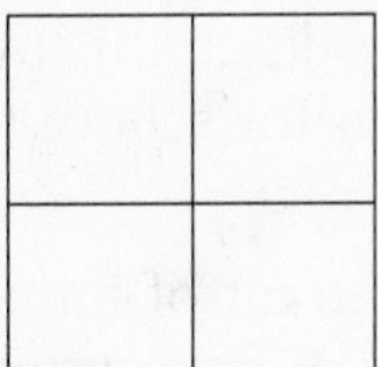

____

3. The Bensons' taxes were increased by 30% over last year. This year their taxes are $2,500. What were last year's taxes?

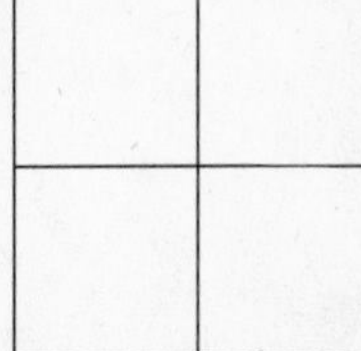

____

LESSON 31

Simple Interest

Simple interest is the cost of using money over a specific length of time.

Borrowed money from a bank, credit union, or mortgage company is a loan. You can get a loan for school tuition, a house, a car, or medical bills.

The **principal** is the amount of money borrowed.
The **interest** is the cost of borrowing the money.
The **rate** is the percent of the principal used to calculate the interest.
The **time** is the length of time of the loan. It is expressed in years.
The formula for finding the cost of a loan is:

Interest = Principal × Rate × Time

Example

Find the interest on a $1,500 loan at 12% for 2 years.

Interest = Principal × Rate × Time

Interest = (1,500 × 0.12) × 2
= (180) × 2
Interest = $360

 For more information, see Book 3, pages 161–162.

Practice

Write the principal, the rate, the interest, and the time for the problems below.

1. The Hague family made a loan for $14,500. The rate was 9%. The simple interest for one year is $1,305.

(1) Principal ______

(2) Rate ______

(3) Interest ______

(4) Length of Time ______

2. Marguerite borrowed $1,600 for tuition. The rate was 5.5% for two years. The interest was $176.

(1) Principal ______

(2) Rate ______

(3) Interest ______

(4) Length of Time ______

Problem Solving

Solve the following problems.

3. What is the interest on $1,600 at 18% for 1 year? ______

4. What is the interest on $2,000 for 6 years at a rate of 7%? ______

5. What is the interest on $500 at a rate of 9% for 3 years? ______

LESSON 32

Ratios

A ratio is a comparison of two numbers. Ratios can be written in three ways:

1. **Fraction Method** $\frac{\text{5 teachers}}{\text{65 students}}$
2. **Ratio Symbol (:)** 5 teachers: 65 students
3. Using **to, per, for, at,** or **in.** 5 teachers per 65 students
   5 teachers for 65 students

MATH HINT

Ratios must always have two numbers. Never write a ratio as a mixed number. Ratios written as fractions can be reduced.

Example

Write the ratio of 24 boxes to 6 crates using the three different methods.

1. **Fraction Method** $\frac{\text{24 boxes}}{\text{6 crates}}$ *or* $\frac{\text{4 boxes}}{\text{1 crate}}$
2. **Ratio Symbol (:)** 24 boxes : 6 crates *or* 4 boxes : 1 crate
3. Using **to, per, for, at,** or **in.** 24 boxes **to** 6 crates *or* 4 boxes to 1 crate

Practice

Use the fraction method to write the following ratios.

1. 15 to 4 ______

2. 9 to 15 ______

3. 18 to 14 ______

4. 16 to 6 ______

5. 14 to 24 ______

6. 12 to 20 ______

 For more information, see Book 3, pages 201–207.

7. 12 quarts to 6 quarts

____

8. 18 pints to 15 pints

____

Use the "ratio symbol (:)" to write the following ratios.

9. 18 miles per gallon

____

10. one inch on a map to 10 miles

____

11. three cups of punch for each person

____

12. 5 yards of cloth per gown

____

13. two cars to one household

____

14. 21 boxes to a crate

____

15. three trees to every five feet

____

16. 3 ounces of cheese to every sandwich

____

17. ten yards of fabric to every sofa cover

____

18. 3 cups of milk to each recipe

____

LESSON 33

Proportions

A **proportion** is a statement that shows two ratios are equal. In a proportion, the **cross products** are **equal**.

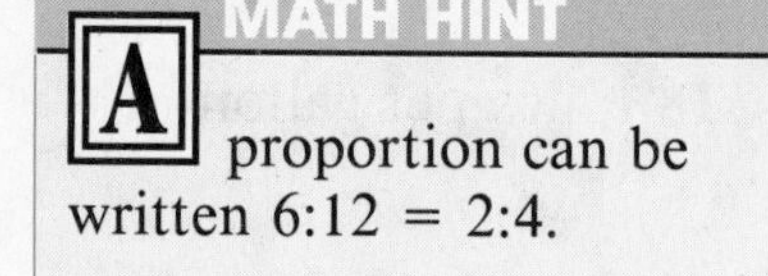

MATH HINT

A proportion can be written 6:12 = 2:4.

For example, $\frac{5}{10} = \frac{1}{2}$

$5 \times 2 = 10$

$10 \times 1 = 10$

To solve a proportion means to find a missing number in the proportion. To find the missing number, follow these steps:

Step 1 Let the letter n stand for the missing number.

Step 2 Find the cross-product of the numbers you know.

Step 3 Divide the cross-product by the number that is left.

Examples

A. Solve the proportion $\frac{3}{4} = \frac{?}{16}$.

Step 1 $\frac{3}{4} = \frac{n}{16}$

Step 2 $3 \times 16 = 48$

Step 3 $\frac{48}{4} = 12 \qquad n = 12$

B. A recipe calls for one-half cup of sugar to make four servings. How much sugar is needed for 12 servings?

Step 1 $\frac{\frac{1}{2}}{4} = \frac{n}{12}$

Step 2 $\frac{1}{2} \times 12 = 6$

Step 3 $\frac{6}{4} = 1\frac{2}{4} = 1\frac{1}{2}$ cups

Practice

Find each missing number.

1. $\frac{5}{8} = \frac{n}{88}$ $\qquad n =$ ______

2. $\frac{5}{n} = \frac{10}{4}$ $\qquad n =$ ______

3. $\frac{n}{48} = \frac{7}{12}$ $\qquad n =$ ______

4. $\frac{11}{3} = \frac{n}{\frac{1}{6}}$ $\qquad n =$ ______

 For more information, see Book 3, pages 208–209; 214–215.

Problem Solving

Solve the following problems.

5. A softball team played 20 games and won 16 of them.

(1) Write the ratio of the number of games won to the number of games lost. ______________

(2) Write the ratio of the number of games lost to the number of games played. ______________

(3) Write the ratio of the number of games won to the number of games played. ______________

6. A small company has 30 employees: 12 men and 18 women.

(1) Write the ratio of men to women. ______________

(2) Write the ratio of women to men. ______________

(3) Write the ratio of men to the total number of employees. ______________

(4) Write the ratio of women to the total number of employees. ______________

LESSON 34

Problem Solving—Proportions

Use the following steps to help you solve proportion word problems.

Step 1 Read the problem and underline the key words for the two things being compared. Write the two things being compared one over the other. Use the variable n is used for the missing number.

Step 2 Make a plan to solve the problem. Solve using cross-products.

Step 3 Find the solution. Use your math knowledge to find your answer.

Step 4 Label your answer.

> **MATH HINT**
>
> Make sure that the two things being compared in a proportion problem remain **in the same order**.

Example

Solve for the proportion in the following word problem.

Margot works as a wardrobe assistant for the circus. She can pack 50 costumes per wardrobe trunk. How many wardrobe trunks will she need to hold 350 costumes?

Step 1 Read the problem and identify the key word, which is **per**. The two things being compared are costumes to wardrobe trunks. Write the proportion in the same order as follows:

$$\frac{\text{costumes}}{\text{wardrobe trunks}} = \frac{50}{1}$$

Step 2 Solve using cross-products.

$$\frac{50}{1} = \frac{350}{n}$$

Step 3 Find the solution.

$$\frac{1 \times 350}{50} = \frac{350}{50} = 7$$

Step 4 Label your answer.

$n = 7$ wardrobe trunks

Margot will need 7 wardrobe trunks to pack 350 costumes.

 For more information, see Book 3, pages 214–215.

Practice

Solve the following problems.

1. Eight ounces of garden fertilizer are mixed with 2.5 gallons of water. How many ounces of fertilizer must be mixed with 10 gallons of water? ____________

2. A 1-gallon can of paint covers 225 square feet. How much area will 3.25 gallons of paint cover? ____________

3. It takes 4 hours to cut 10 acres of lawn. How long will it take to cut 45 acres? ____________

4. On a city street map, 1.5 inches represent 1 mile. How many miles is a road that measures 12.75 inches on the map? ____________

5. One dozen doughnuts cost $1.80. How much do 22 doughnuts cost? ____________

UNIT

4

Posttest

Change each decimal to a percent.

1. 0.362 ______

2. 0.63 ______

3. 1.45 ______

Change each percent to a decimal.

4. 17% ______

5. 145% ______

6. 4.5% ______

Change each fraction to a percent.

7. $\frac{1}{48}$ ______

8. $\frac{2}{39}$ ______

9. $\frac{5}{8}$ ______

Change each percent to a fraction.

10. 40% ______

11. 114% ______

12. 55.5% ______

Solve the following proportion problems.

13. $\frac{3}{7} = \frac{21}{n}$

$n =$ __________

14. $\frac{4}{n} = \frac{10}{8}$

$n =$ __________

15. $\frac{n}{36} = \frac{7}{12}$

$n =$ __________

Write each ratio as a fraction.

16. 12 to 7 ________

17. 35 to 21 ________

18. 38 to 14 ________

19. 10 pints to 3 pints ________

20. 6 quarts to 5 gallons ________

Problem Solving

Solve the following problems.

21. A large bread machine uses 3 cups of flour to make a loaf of bread that weighs $1\frac{1}{2}$ pounds. How many cups of flour are needed to make a loaf that weighs 2 pounds?

____

22. A team plays 36 games. If the team won 30 games, what is the ratio of games won to games lost?

____

23. A $19,000 boat loses 25% of its value in depreciation in the first year. How much does the boat depreciate?

____

24. On the opening night of a movie, a theater sold all 500 seats. On the second night, 425 tickets were sold for the movie. By what percent did sales decrease?

____

25. The asking price for a used motorcycle was decreased from $1,800 to $1,500. What was the percent of decrease?

____

26. The price of a coffee maker was increased from $35 to $42. What was the percent of increase in price?

____

UNIT

5

Graphs, Measurements, and Statistics

Pretest

Refer to the graph below to answer questions 1–4.

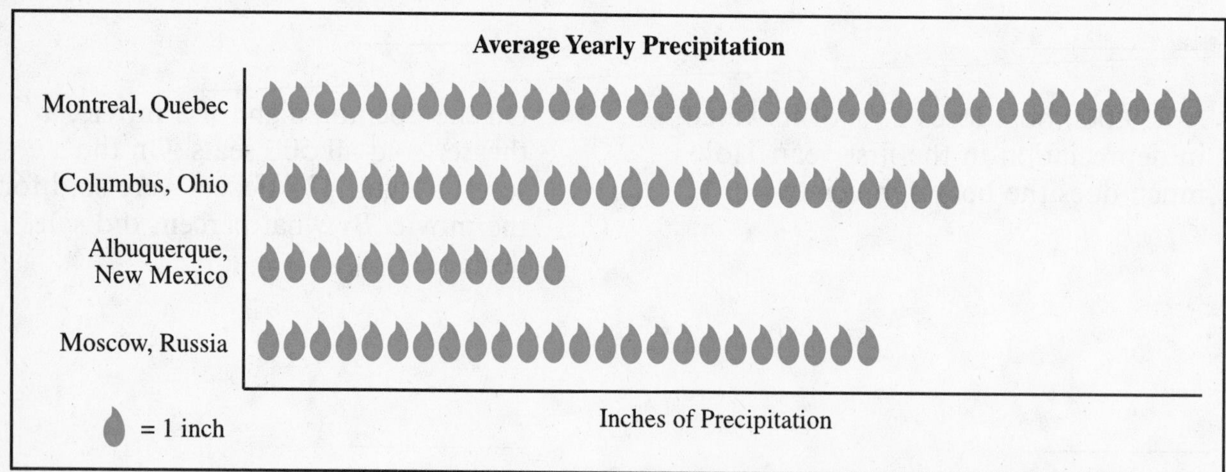

1. Which city has the highest average yearly precipitation?

____

2. Which city has the lowest yearly precipitation?

____

3. What is the average yearly precipitation for Moscow?

____

4. How much more average yearly precipitation does Columbus, Ohio, receive than Albuquerque, New Mexico?

____

Use the graph below to answer questions 5–8.

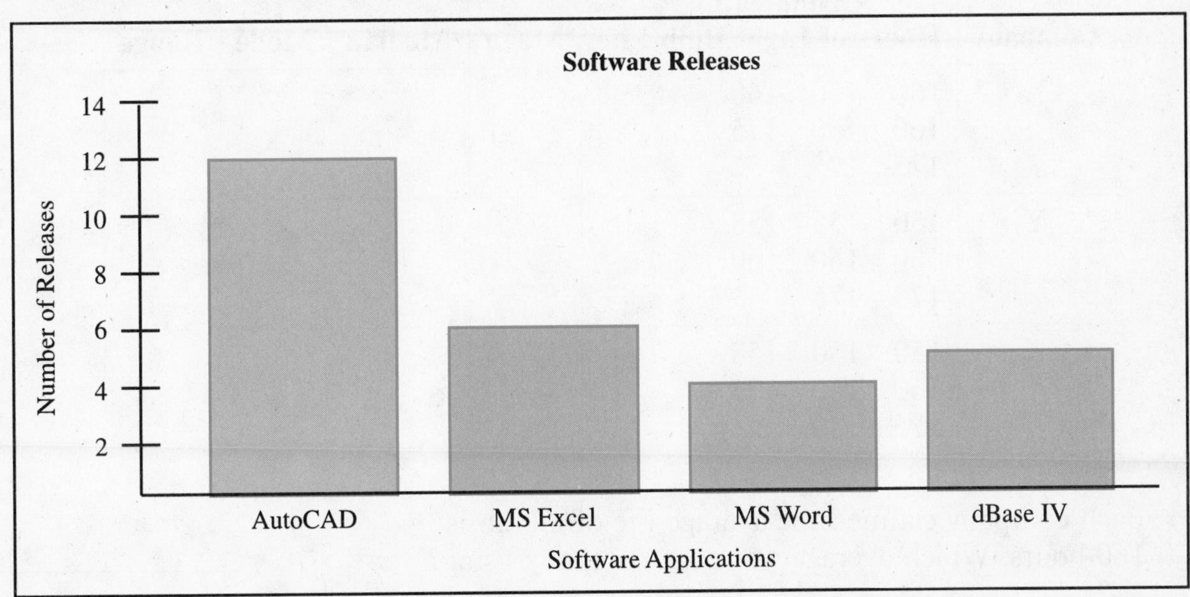

5. Which software application has the smallest number of releases?

____

6. Which software application has the largest number of releases?

____

7. How many more releases has AutoCAD than MS Word?

____

8. How many more releases has MS Excel than MS Word?

____

Problem Solving

Complete the table below. Then solve the problem.

9. Three light bulb manufacturers each claim that the average life of their 60-watt bulb is 160 hours.

Company	Number of Hours of Light Bulb Life	Mean	Median	Mode	Range
X	160 160 160 160 160 175 175 175				
Y	150 155 158 160 160 160 175 175				
Z	150 150 158 160 160 175 175 180				

(1) Each company claimed the average life of a bulb is 160 hours. Which average was each company using? ____________

(2) Which company seems to have the best light bulb? ____________

Graphs

Graphs show how several pieces of information compare. The graphs used most often are **circle**, **bar**, **line**, and **picture** graphs.

Examples

A **circle graph** shows how a total amount is divided into parts. The entire circle represents 100% of the total data. The parts of the graph relate to the whole.

Look at the circle graph below. The entire circle represents 100% of the world's oceans. The ocean is the entire body of saltwater that covers the earth's surface. This body of water is divided into four parts: the Pacific Ocean, the Atlantic Ocean, the Indian Ocean, and the Arctic Ocean. The circle graph compares the sizes of the world's oceans by the sizes of its parts.

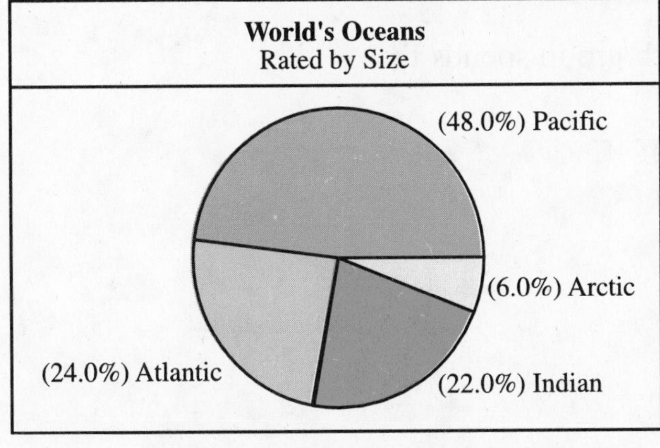

The part of the circle standing for the size of the Atlantic Ocean (24.0%) is half the size of the part standing for the Pacific Ocean (48.0%). This is because the Atlantic Ocean is about half the size of the Pacific Ocean.

A. According to the circle graph, which ocean is the smallest? Which two oceans are about the same size?

The Arctic Ocean is the smallest. It makes up 6.0% of the world's oceans.

The Indian Ocean (22.0%) and the Atlantic Ocean (24.0%) are about the same size.

For more information, see Book 4, pages 71–82.

A **bar graph** is useful for organizing and displaying data and for showing how sets of data compare. Bar graphs have either **horizontal** or **vertical** bars.

The bar graph below shows the average cost of home telephones per year by the age of the users. It is easy to see the comparison between age groups.

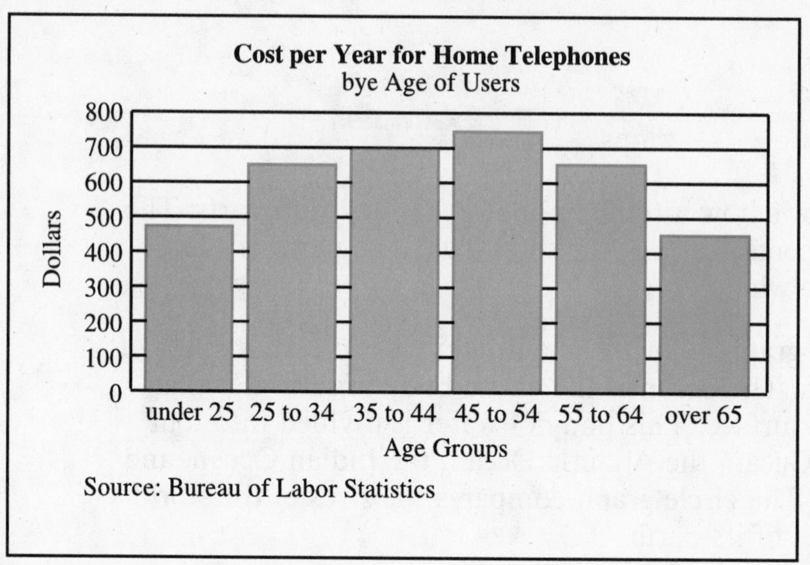

B. Which group spends the most per year? Which group spends the least per year?
The age group 45 to 54 spends the most per year.
The age group over 65 spends the least.

A **line graph** shows changes in the relationship between two items. The horizontal and vertical axes may be scaled differently. A line graph is useful to show trends or changes over a period of time. The line graph below shows the average SAT scores of high school students in the United States between 1960 and 1990.

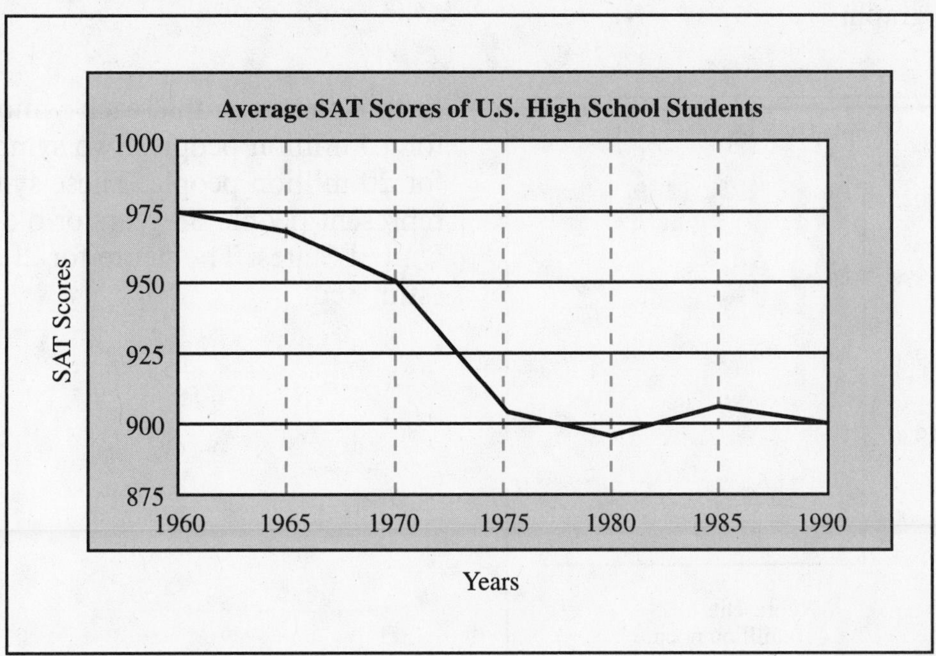

C. Find the average score for the high school students in 1970.

To find this score, first locate 1970 on the horizontal axis. Move up to the line graph and then read over on the vertical axis to find the score. This score is about 950.

A **picture graph** shows data by having a symbol represent a given amount. By comparing the data, you can determine important information. Picture graphs show contrast and emphasize ideas.

To read a picture graph, you need to know what each picture represents. Then you count the symbols to find the value of the data shown on the graph.

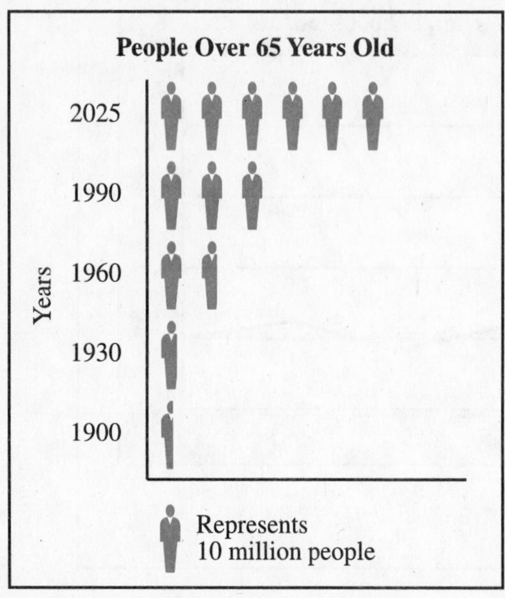

In this picture graph, each symbol stands for 10 million people. Two symbols stand for 20 million people. These symbols represent people 65 years or older in the United States. The figure for 2025 was estimated.

D. How many people in the United States were over the age of 65 in 1930?

About 75% of the symbol is shown. This means that 75% of what the symbol represents, or 7.5 million people, were over 65 in 1930.

Practice

Use the circle graph below to answer problems 1–4.

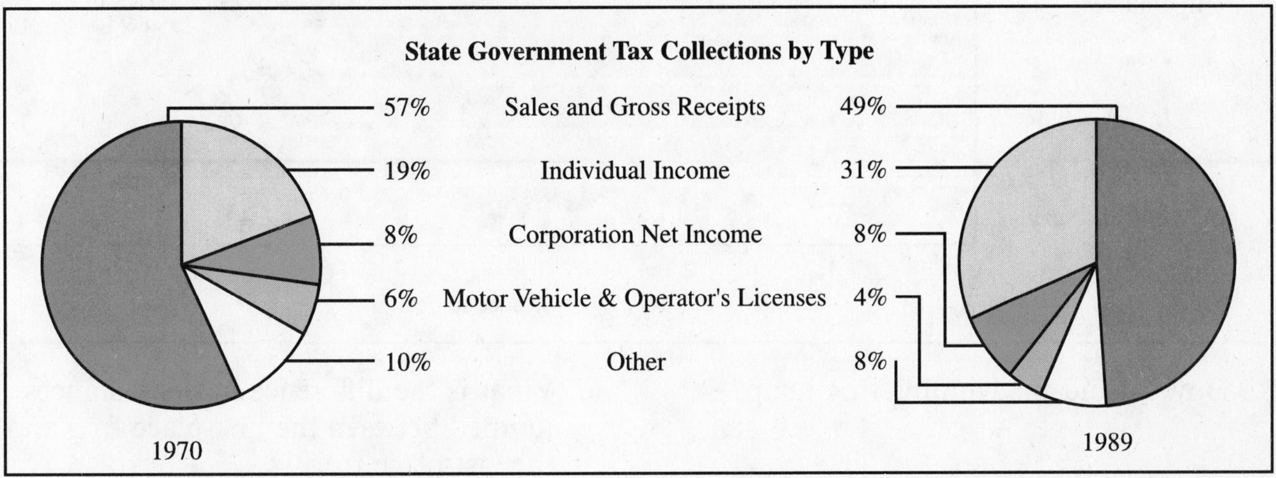

1. Which group made up the highest percent of tax collections for 1970?

____

2. Which group made up the lowest percent of tax collections for 1989?

____

3. Which category shows the greatest decrease in the percent of tax collected between 1970 and 1989?

____

4. In which category would estate taxes be included?

____

Use the bar graph below to answer problems 5–8.

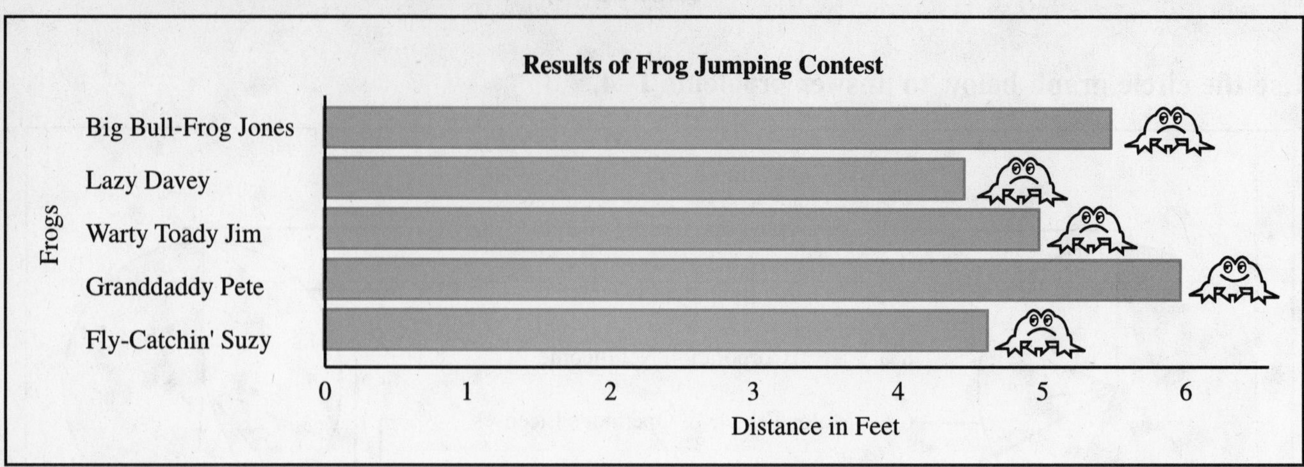

5. How far did the winning frog jump?

____

6. What is the difference in the distances jumped between the first-place frog and the last-place frog?

____

7. How far did Warty Toady Jim jump?

____

8. How many frogs jumped more than 5 feet?

____

Use the bar graph below to answer problems 9–14.

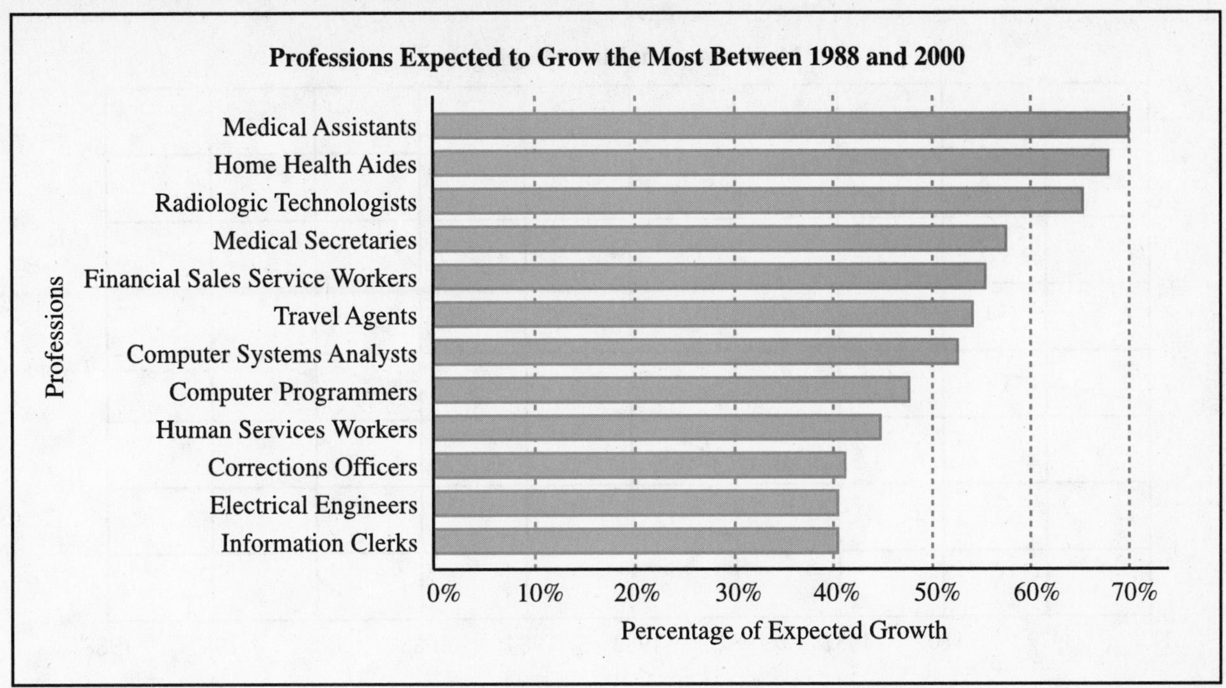

9. What is the percent of expected growth for medical secretaries between 1988 and 2000?

____

10. What is the percent of expected growth in human services workers between the years 1988 and 2000?

____

11. Of the 12 categories listed, which profession is expected to grow the most by 2000?

____

12. Which professions are expected to grow more than 60% by 2000?

____

13. Which two professions are expected to grow equally?

____

14. Suppose there were 58 computer programmers in Estatesville in 1988. Use the graph to estimate the number of programmers Estatesville will need in the year 2000.

____

Use the line graph below to answer problems 15–18.

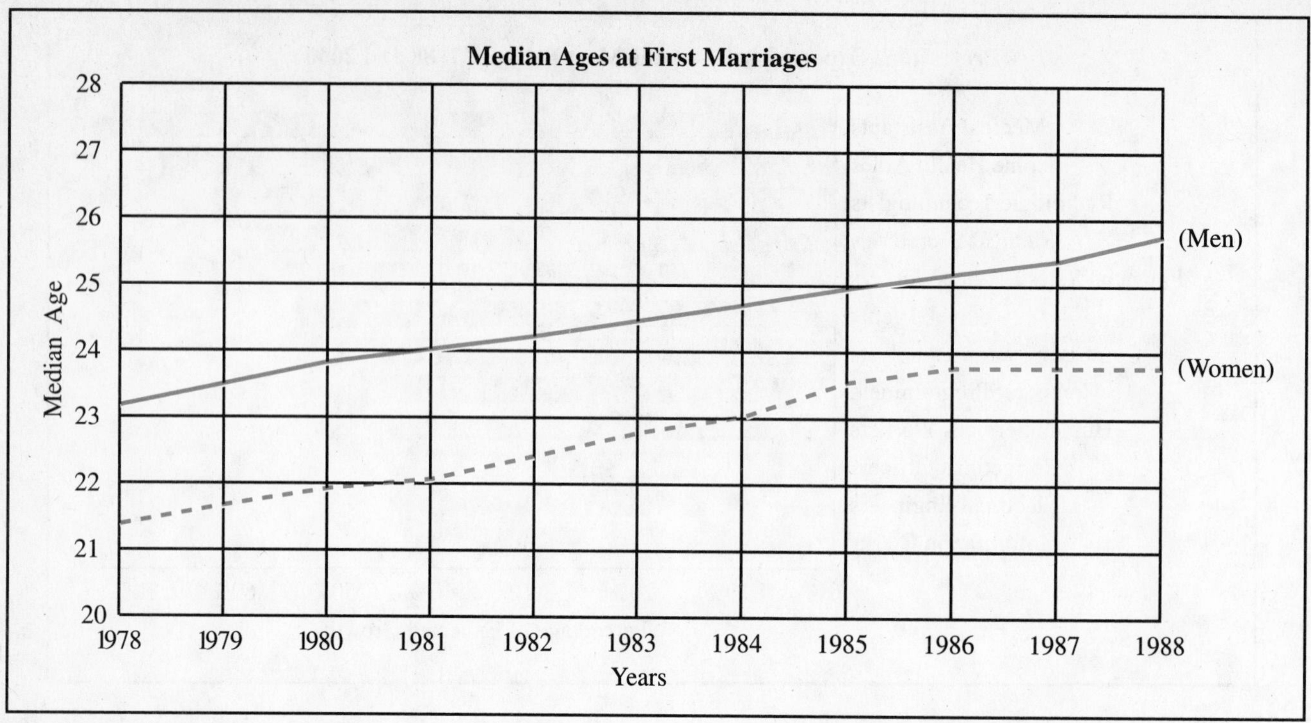

15. What was the median age of women getting married for the first time in 1981?

____

16. Estimate the median age increase for men getting married for the first time between 1978 and 1988.

____

17. Which year had the lowest median age for marriage for both sexes?

____

18. In which year was the median age the highest for men?

____

LIFE SKILL

Interpreting Graphs

Graphs provide us with a picture of what is happening. It is important to read and interpret graphs. As you look at graphs, think about what is happening.

This line graph shows the level of the water in a bathtub. Three periods are marked off: the period of time when the tub is filling; the period when the tub is filled; and the period when the tub is emptying.

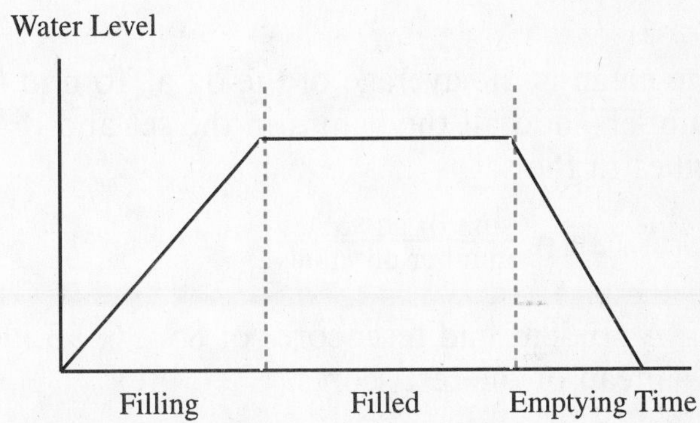

Professionals such as physicians and engineers use graphs to help them understand what is happening. Physicians may use graphs to study how a patient's heart functions; engineers use them to understand how materials react under certain stresses.

Can you guess what is happening by looking at graphs? Match each graph to the description that fits best.

1. The speed of a car on a city block between stop signs.

____

A.

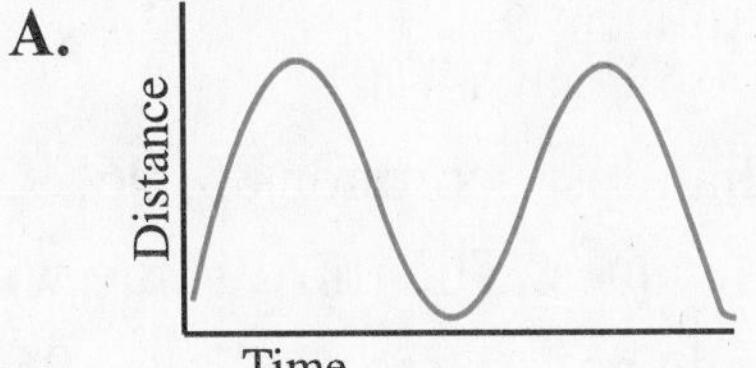

2. The speed of a Velcro® ball flying through the air and hitting a Velcro® wall.

____

B.

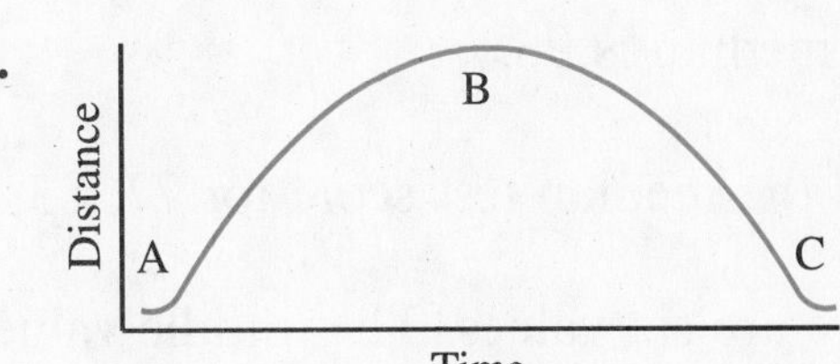

3. The distance between the floor and a yo-yo going up and down.

____

C.

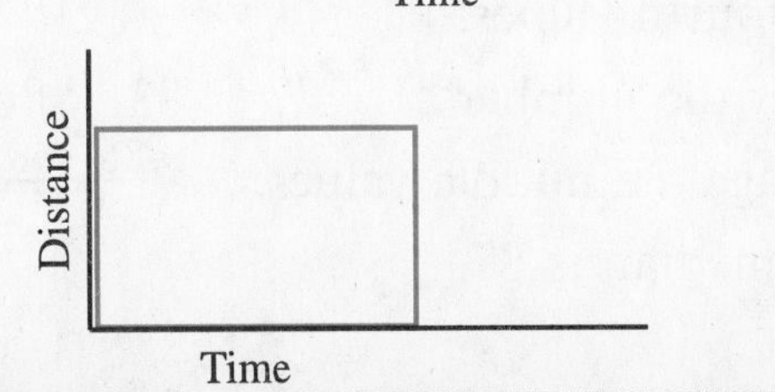

Statistics

Numbers known as **measures of central tendency** represent the middle values, or the center, of a set of data. Statisticians often use these numbers to analyze data. The three most common measures of central tendency are the **mean**, the **median**, and the **mode**.

Examples

The **mean** is the average of the data. To find the mean of a set of numbers, add all the values in the set and divide by the number of values in the set.

$$\text{Mean} = \frac{\text{sum of all values}}{\text{number of values}}$$

A. A student had test scores of 85, 96, 93, 96, and 100. Find the mean of the test scores.

Add all the values together. $85 + 96 + 93 + 96 + 100 = 470$

Divide by the number of values. $\frac{470}{5} = 94$

The mean is 94.

The **median** is the middle value in a set of ordered data. To find the median, arrange the numbers in order. For an odd number of values in the list, the middle number is the median. For an even number of values in the list, find the average of the two middle numbers. This number is the median.

B. Find the median test score for 85, 96, 93, 96, and 100.

There are five values. The third value will be the median.

Order the numbers. 85 93 96 96 100

The median is 96.

C. Find the median test score for 77, 83, 92, 94, 86, and 88.

There are six values. The middle values are the third and fourth values.

Order the numbers. 77 83 86 88 92 94

Average the middle values. $\frac{86 + 88}{2} = \frac{174}{2} = 87$

The median is 87.

MATH HINT

The median strip on a highway is the middle of the highway. Use this hint to help you remember that the median is the middle number.

 For more information, see Book 4, pages 135–149.

The **mode** of a set of numbers is the number or numbers that occur most often. To find the mode, count the times a number occurs.

D. A student had test scores of 85, 96, 93, 96, and 100. Find the mode.

Value	**Number of Occurrences**
85	1
93	1
96	2
100	1

The mode is 96.

MATH HINT

There can be more than one mode.
1, 3, 4, 4, 4, 5, 6, 6, 6, 8, 9
This set of data has two modes: 4 and 6.

The **range** of a set of numbers is the difference between the highest value and the lowest value.

E. A student had test scores of 85, 96, 93, 96, and 100. Find the range of the test scores.

The highest value is 100 and the lowest value is 85.
Subtract these values to find the range.

$$100 - 85 = 15$$

The range is 15.

Practice

Solve the following problems.

1. Complete the table below.

Values	Mean	Median	Mode	Range
28 32 33 35 42				
21 25 28 49 53				

2. Complete the table below. Answer the questions that follow.

Company	Number of Hours of Battery Life	Mean	Median	Mode	Range
X	14 14 14 14 16 16 16 16				
Y	11 12 13 15 15 15 16 17				
Z	11 11 12 13 13 14 18 19				

Each company claims that the average life of its battery is better than that of the other manufacturers. Which battery do you think has the longest life? Why do you think so?

____

____

3. The Fly by Night Agency has 8 employees: 4 men and 4 women. The women feel that the pay system discriminates against them. Complete the table below. Answer the questions that follow.

Gender	Salary	Mean	Median	Mode	Range
Women	\$60,000 \$30,000 \$25,000 \$25,000				
Men	\$35,000 \$35,000 \$35,000 \$35,000				

(1) If you were an arbitrator hired to represent the women, which types of averages would you use to argue the case for the women?

____

(2) Which types of average show the men's average salaries higher than the women's average salaries?

____

Problem Solving—Making a Table

The steps you have learned to help solve word problems can be used with word problems that deal with statistics. Use the following steps:

Step 1 Read the problem and underline the key words. These words will usually relate to some mathematical reasoning.

Step 2 Make a plan to solve the problem. Ask yourself, Should I add, subtract, multiply, divide, round, or compare? You may have to do more than one of these operations for the same problem.

Step 3 Find the solution. Use your math knowledge to find your answer.

Step 4 Check your answer. Ask yourself, Is this answer reasonable? Did you find what you were asked for?

When solving a word problem, often you can make a table to help you organize information.

Example

The owner of a leather store had a booth at an exhibition of local artists and craftsmen. In one weekend, he sold the following sizes of tooled leather belts:

32	36	34	42	46	34
40	44	38	42	36	42
32	46	38	36	42	42
40	38	36	36	42	40

Find the mean, median, and mode of the belt sizes sold.

Step 1 Read the problem and underline the key words: **mean**, **median**, and **mode.**

Step 2 Make a plan to solve the problem. Ask yourself these questions:

What do you want to find? mean, median, mode

What do you know? list of sizes sold

How do you find what you need?

For more information, see Book 4, pages 135–149.

mean—add sizes together and divide by 24 (the total number of belts)

median—order the numbers and pick the middle number

mode—order the numbers and pick the size that is listed most often

Step 3 Find the solution. Create a table to organize the data.

Size	Number of Belts Sold	Size	Number of Belts Sold
32	2	40	3
34	2	42	6
36	5	44	1
38	3	46	2

The mean is the sum of the sizes divided by 24.

$$\frac{934}{24} = 38.91666$$

The mean is about 38.9.

The median is the average of the 12th and 13th values. The 12th item is 38. The 13th item is 40.

$$\frac{38 + 40}{2} = 78$$

The median is 39.

The belt sold most often is size 42.
The size 42 has 6 items sold.
The mode is 42.

Step 4 Check your answer.

$24 \times 38.9 = 933.6$ Checking the sum, you get 934.

This checks. By definition, the median of 39 and mode of 42 check.

Practice

Complete the table for each problem. Then solve the problem.

1. A student needs a mean score of 80 for 11 tests in her mathematics course. Her mean score of 10 tests for the term is 79. How many more points does she need to have a mean of 80?

   ____

Test Score	Number of Times Earned	Sum of Scores
79	10	
x	1	
Totals	11	

2. A student needs a mean of 80 to pass a course. On the first four tests, he earned the following scores: 70, 86, 88, and 76. What is the lowest grade he can earn on the fifth test and still have a mean of 80?

   ____

Test Score	Number of Times Earned	Sum of Scores
70	1	
76	1	
86	1	
88	1	
x	1	
Totals	5	

3. The Reynolds family has an annual income of $30,000 and has kept records of their expenses for a year. The table below shows the percents for each category. Find the amount spent on each category.

Category	Percent	Amount Spent
Housing	25%	
Food	20%	
Transportation	15%	
Clothing	12%	
Medical	13%	
Recreation	10%	
Savings	5%	
Total	100%	

UNIT

5

Posttest

Refer to the graph below to answer questions 1–4.

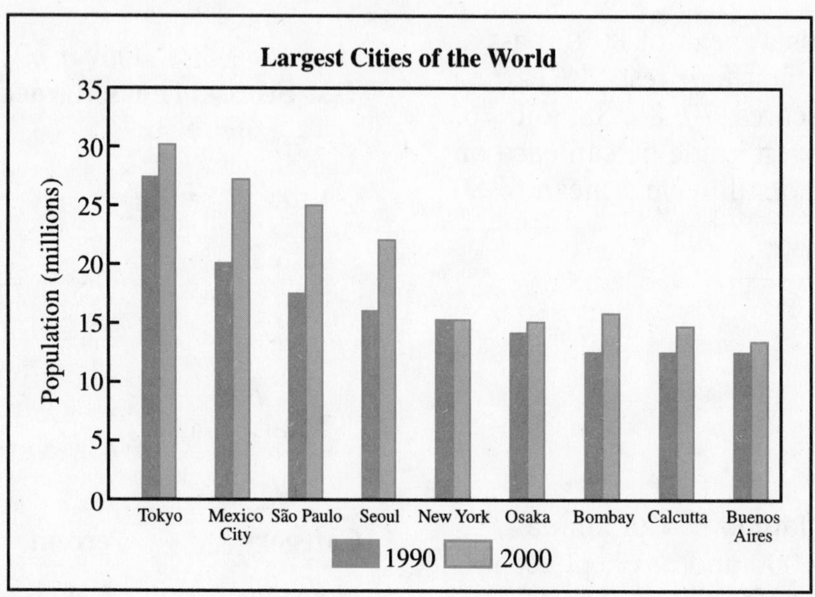

1. Which city has the largest population projected for the year 2000?

____

2. Which two cities have the smallest projected change in population?

____

____

3. Which two cities have the largest projected change in population?

____

____

4. What is the projected population of Bombay in the year 2000?

____

Use the graph below to answer questions 5–8.

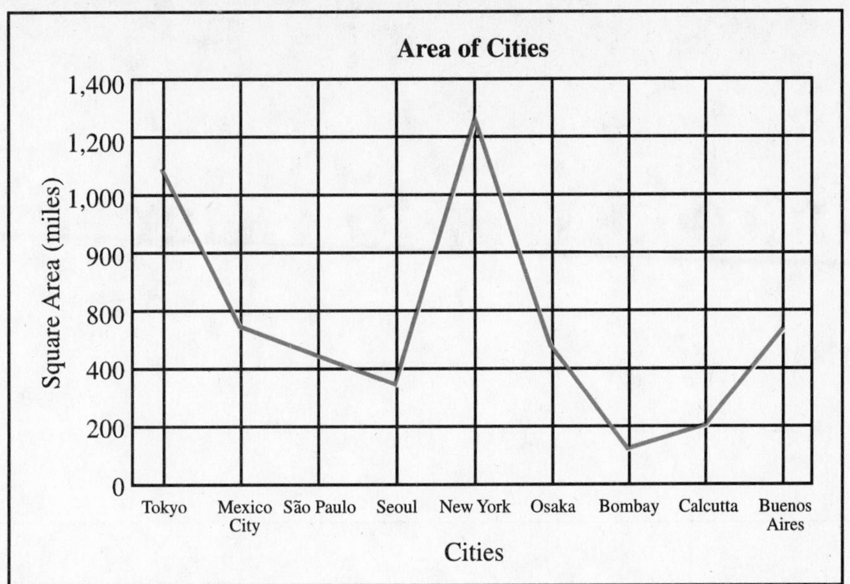

5. Which city has the largest area?

____

6. Which city has the smallest area?

____

7. What is the difference between the area of Seoul and that of Mexico City?

____

8. Using the bar graph on population, which city is probably the most crowded per square area?

____

Use the data listed below to answer questions 9–13.

14	14	14	14	16	16	16	16	11	12
16	17	11	11	12	13	13	14	18	19

9. Find the mean.

____

10. Find the mode.

____

11. Find the median.

____

12. Find the range.

____

13. Which average would you use to best describe the average value?

____

UNIT

6

Introduction to Geometry

Pretest

Identify each of the following figures.

1. 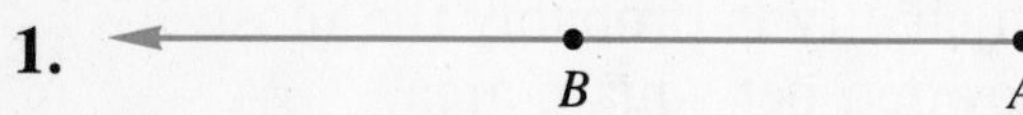

2. 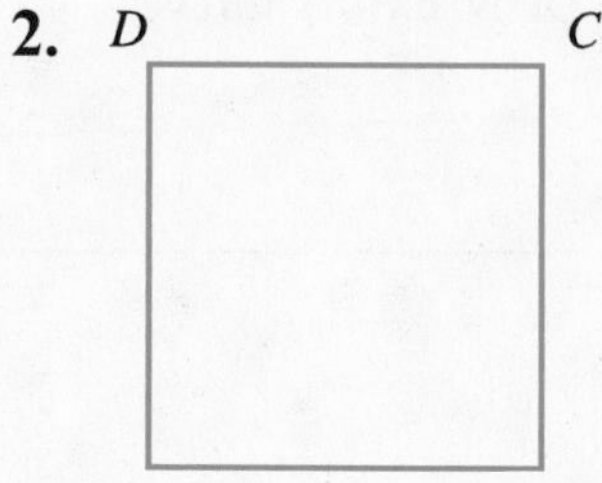

3. D C A B

4. 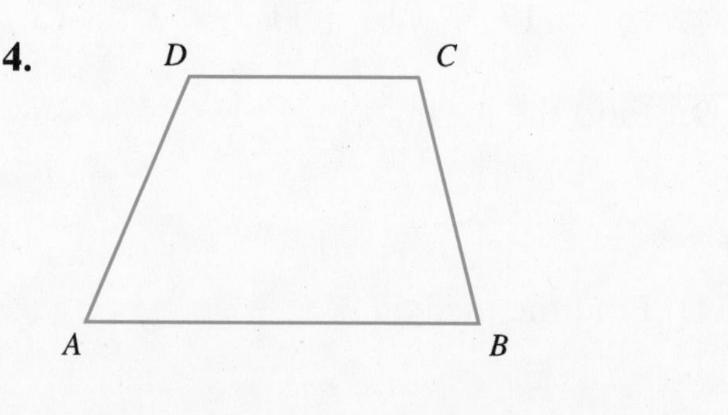

Use the figure below to answer questions 5–8.

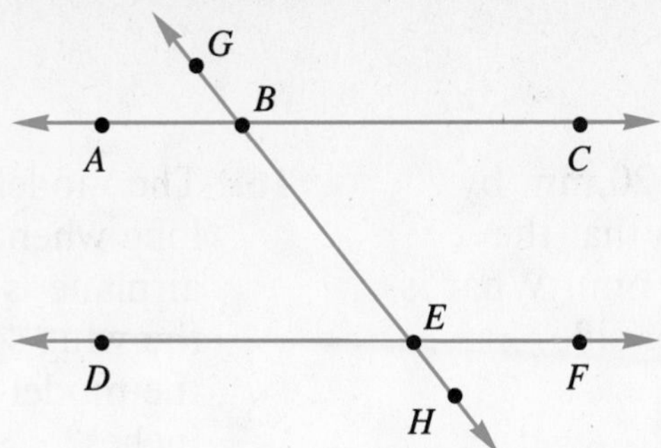

5. Name an acute angle. __________

6. Name an obtuse angle. __________

7. Name two pairs of corresponding angles. __________

8. Name two pairs of vertical angles. __________

9. Name a pair of alternate exterior angles. __________

10. Name a pair of alternate interior angles. __________

Describe the following triangles.

11. right triangle __________

____

12. acute triangle __________

____

Find the perimeter and the area for each of the following figures.

13.

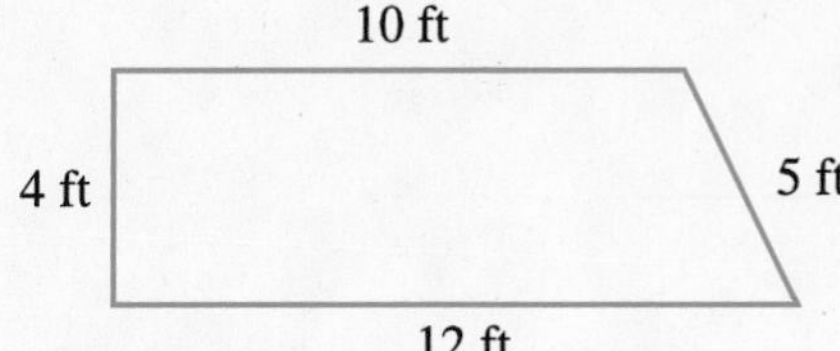

____

____

14.

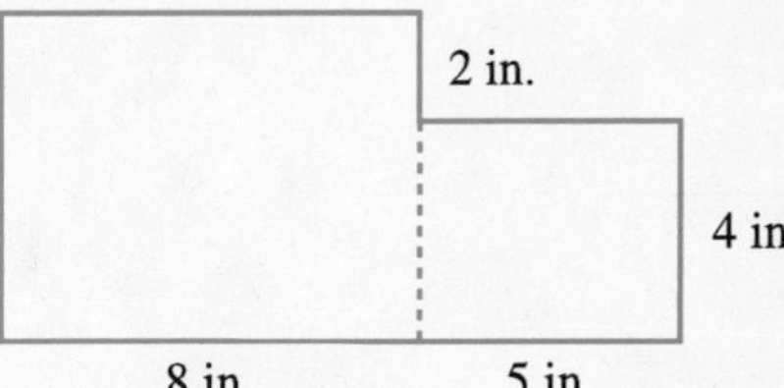

____

____

Problem Solving

Solve the following problems.

15. A photograph measures 120 mm by 150 mm. It is enlarged so that the longer side measures 300 mm. What is the measure of the shorter side?

______

16. The model of an airplane is 24 inches long when the actual length of the airplane is 24 feet long. How wide is the wing span of the actual airplane if the model has a wing span of 36 inches?

______

Find the length of the diagonals for each of the following.

17.

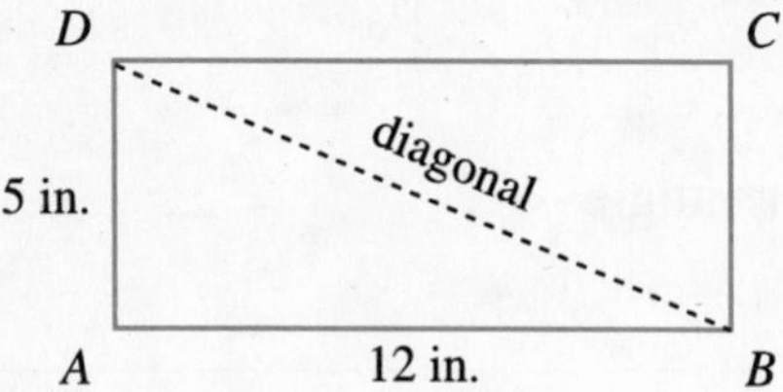

______

18.

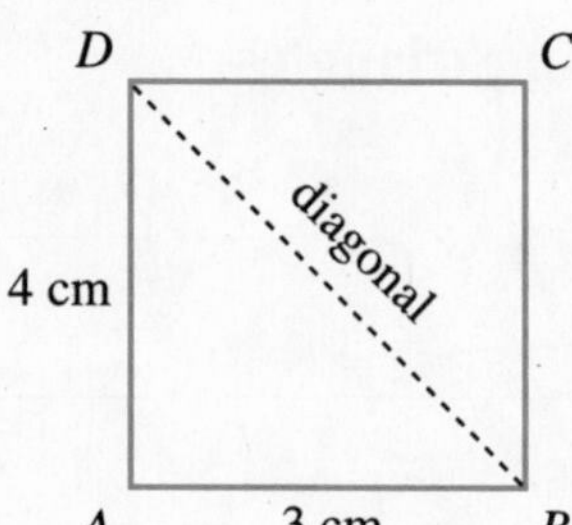

______

Introduction to Geometry

The word **geometry** means "earth measure." In geometry, you study size, shape, and position of objects in space.

Plane geometry is the study of flat surfaces. Plane figures have two dimensions: length and width. The figures shown below are plane figures.

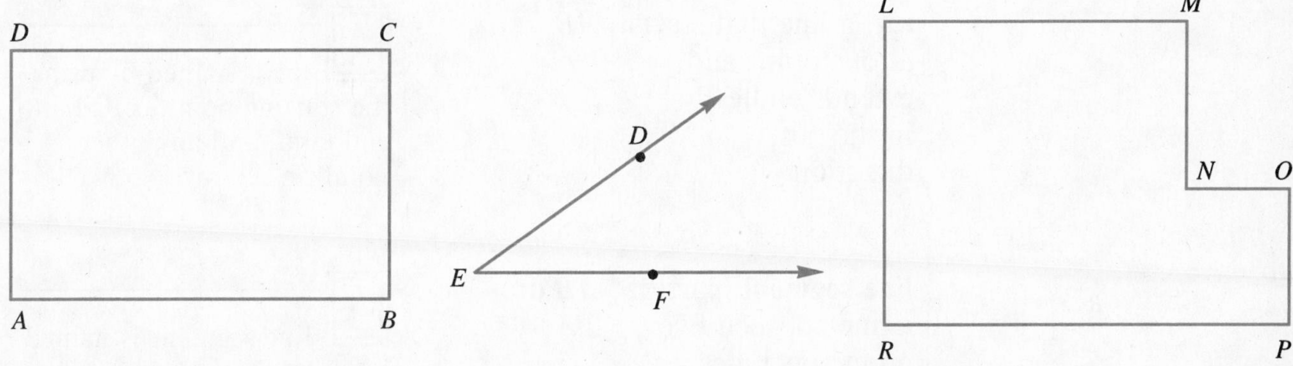

Figures in **solid geometry** have three dimensions: length, width, and height. A baseball, a sugar cube, and a box are examples of solid geometry figures.

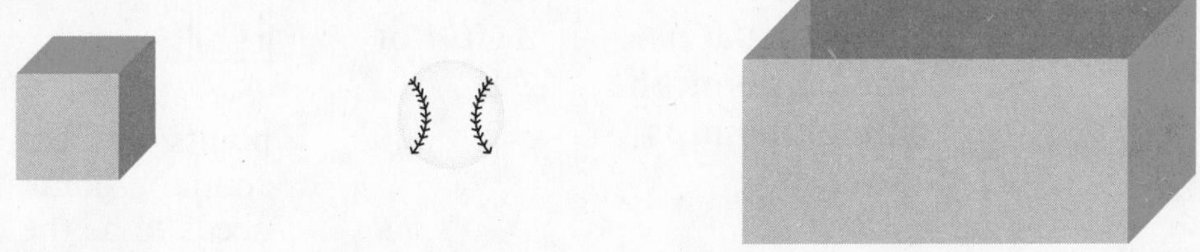

It is necessary to start your study of plane geometry with some definitions. In the chart on the next page, a description and a symbol explain each figure.

MATH HINT

The terms and definitions used in geometry are carefully defined and may be different from the way in which these same terms are used in common speech.

For more information, see Book 5, pages 34–37.

Figure	Description	Symbol
A	**point:** a location in space; has no dimension	A
l, A, B	**line:** a direction in space; extends endlessly in opposite directions	$\overleftrightarrow{AB}$ or line l
A, B	**ray:** a line that starts at one point and extends endlessly in the other direction	$\overrightarrow{AB}$
A, B	**line segment:** part of a line between two points; includes the two points	$\overline{AB}$ or $\overline{BA}$
C, B, A	**angle:** two rays intersecting at a common point; the common point is the vertex	$\angle ABC$ or $\angle CBA$ or $\angle B$

MATH HINT

A line is named by using any two points on the line or a script letter.

MATH HINT

A ray is named by using the starting point, called an endpoint, and any other point on the ray.

MATH HINT

A line segment is named by using the endpoints of the segment.

MATH HINT

An angle is named in several ways. When you use points on the rays and the common point, the vertex needs to be the center letter. When you use the vertex alone, make sure it is clear which angle you are naming.

In a plane, two lines can meet or not meet. If the lines meet at a point, they are intersecting lines.

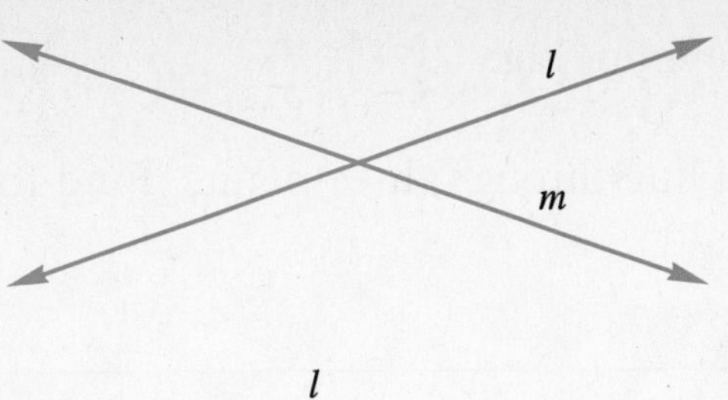

If two lines do not meet even if they are extended, they are parallel lines.

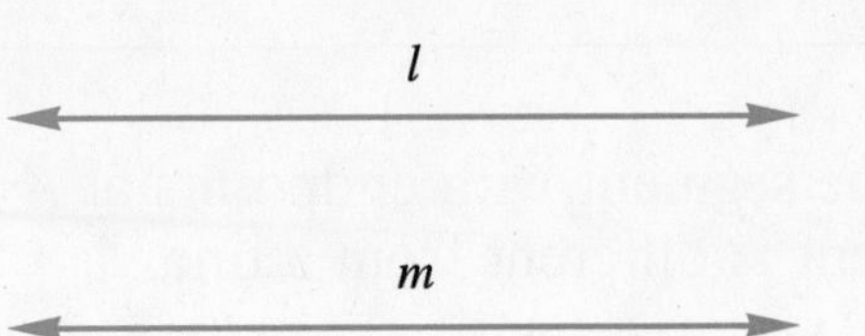

Practice

Identify each of the figures below.

1.

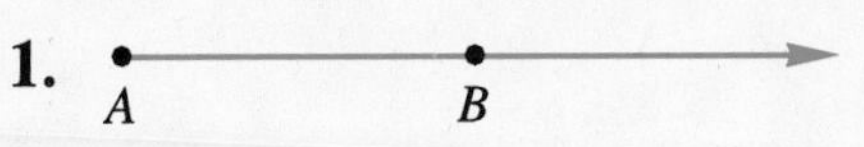

____

2.

____

3.

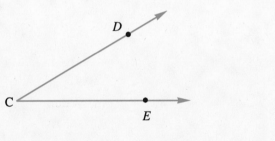

D

C

E

____

4.

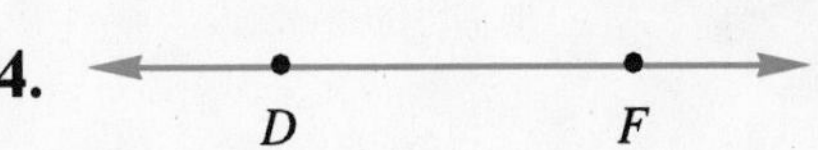

____

5.

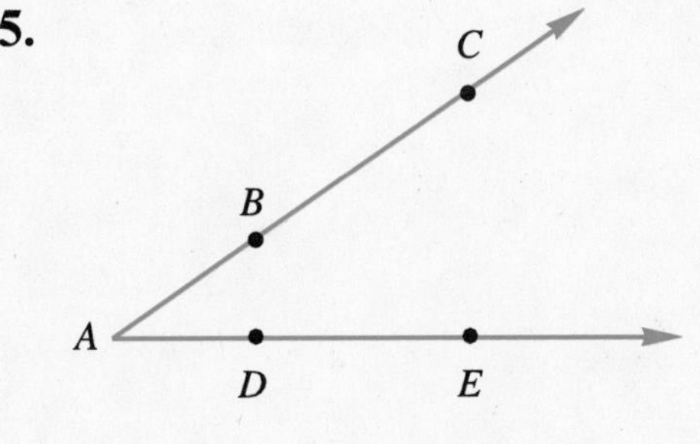

____

6.

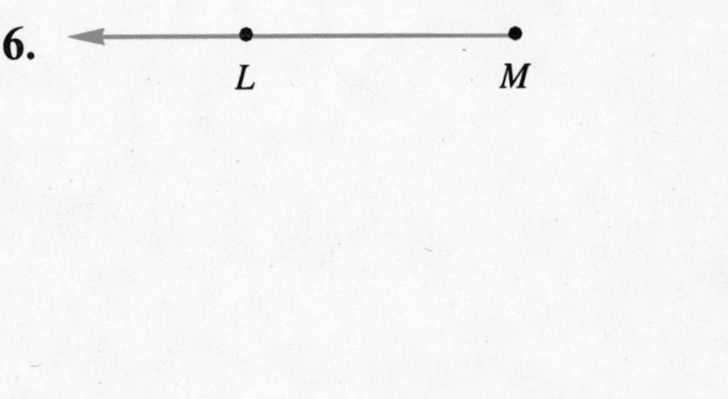

____

Answer the following.

7. Draw a line through three points. Find four ways to name this line.

____

8. Draw a line segment with endpoints at *A* and *B*. Explain how a line segment is different from a line.

____

9. Draw a ray with its endpoint at *B*. Explain how a ray is different from a line.

____

Angles

An **angle** is formed by two rays with a common point. The common point becomes the vertex of the angle. The figure below shows $\overrightarrow{AC}$ and $\overrightarrow{AB}$ meeting at the common point *A*.

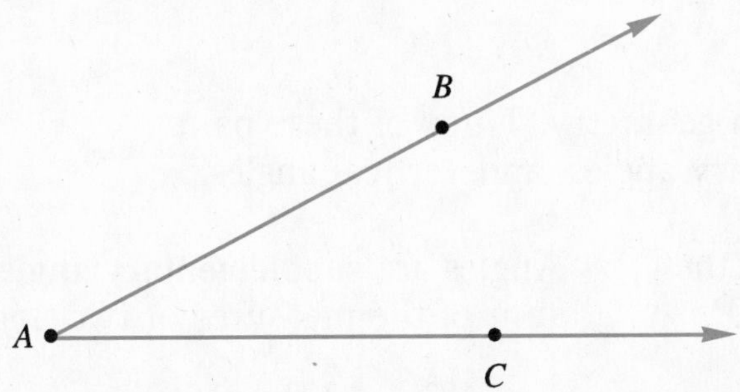

The symbol for angle is ∠. The angle can be named $\angle BAC$, $\angle CAB$, or $\angle A$.

The measure of the angle is the size of the opening between the rays. Angles can be measured in **degrees** (°). You can use a protractor to measure angles. Marks on the protractor represent degrees.

Acute angles have a measure less than 90°.

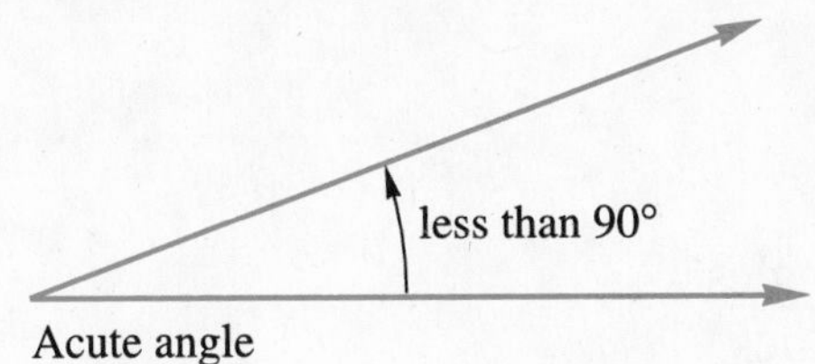

Acute angle

Right angles have a measure equal to 90°. You can mark a right angle by placing a box at the vertex of the angle.

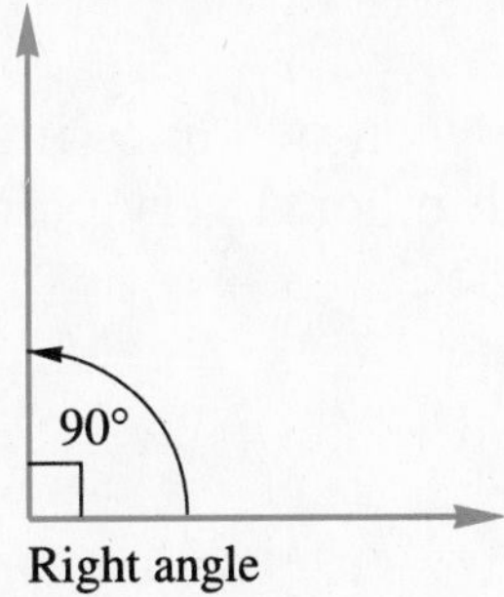

Right angle

Obtuse angles have a measure greater than 90° and less than 180°.

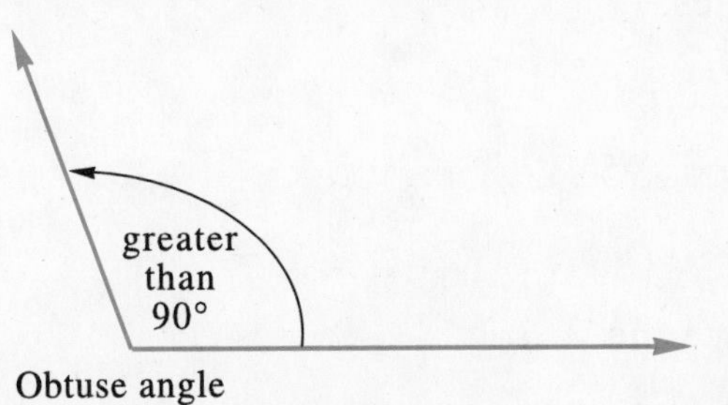

Obtuse angle

A **straight line** has a measure equal to 180°.

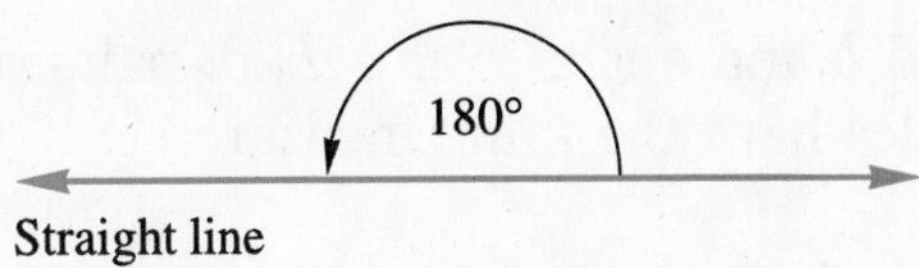

Straight line

For more information, see Book 5, pages 38–44.

Examples

A. What kind of angle is $\angle DCF$?

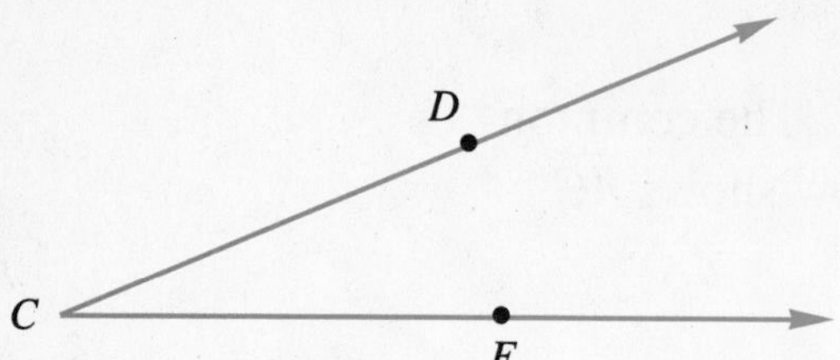

The angle has a measure less than 90°. $\angle DCF$ is an acute angle.

Certain pairs of angles are important in geometry. Three of these pairs are complementary angles, supplementary angles, and vertical angles.

Angles are **complementary angles** when the sum of the measures of two angles is 90°.

$$50° + 40° = 90°$$

Angles with measurements of 40° and 50° are complementary angles.

Angles are **supplementary angles** when the sum of the measures of two angles is 180°.

$$48° + 132° = 180°$$

Angles with measurements of 48° and 132° are supplementary angles.

B. $\angle a$ and $\angle b$ are supplementary angles, and $\angle a = 135°$. Find the measurement of $\angle b$.

Since $\angle a$ and $\angle b$ are supplementary angles, $\angle a + \angle b = 180°$.

By substitution, $135° + \angle b = 180°$

$$\angle b = 180° - 135°$$

$$\angle b = 45°$$

Vertical angles are formed when two lines intersect. Vertical angles are the angles opposite each other. Every pair of intersecting lines has two sets of vertical angles.

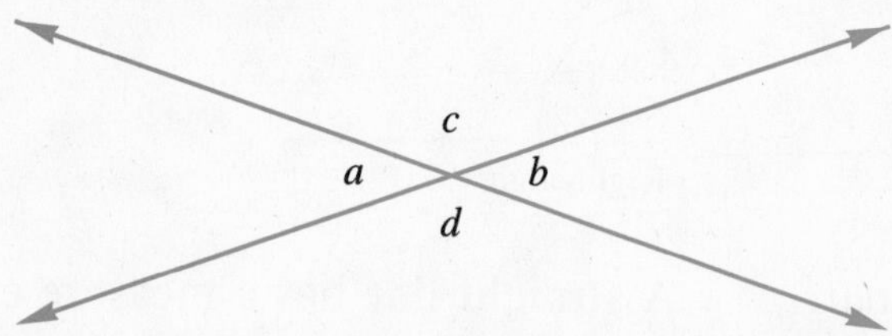

Angles a and b and angles c and d are pairs of vertical angles. Vertical angles have the same measure.

C. $\angle c$ and $\angle d$ are vertical angles. The measurement of $\angle c$ is 95°. Find the measurement of $\angle d$.

The measurement of $\angle d$ is 95° since vertical angles have the same measure.

Parallel lines are lines in the same plane that never meet. The lines are the same distance from each other at every point on the lines.

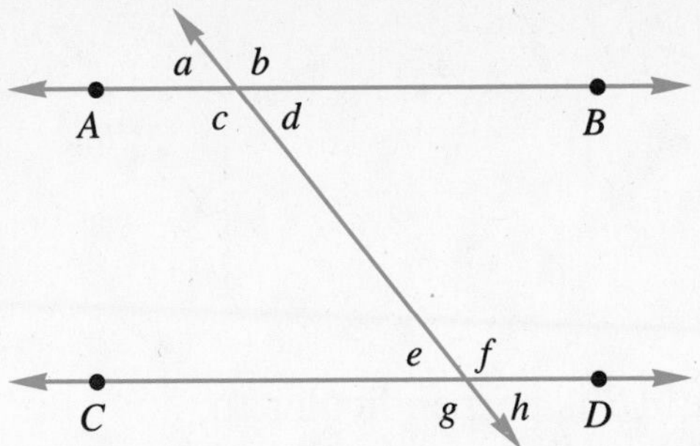

Lines *AB* and *CD* are parallel. This is written $\overleftrightarrow{AB} \parallel \overleftrightarrow{CD}$. It is read "line *AB* is parallel to line *CD*."

A line that intersects two parallel lines is a **transversal**. When a transversal cuts two lines, several angles are formed.

The angles formed and located in the same position compared to the transversal are **corresponding angles**. Corresponding angles are equal. In the figure above, $\angle a$ corresponds to $\angle e$, $\angle b$ corresponds to $\angle f$, $\angle c$ corresponds to $\angle g$, and $\angle d$ corresponds to $\angle h$.

Angles on the outside of the parallel lines are **exterior angles**. Alternating exterior angles are equal. In the figure above, $\angle b$ and $\angle g$ form a pair of alternating exterior angles, and $\angle a$ and $\angle h$ form another pair.

Angles on the inside of the parallel lines are **interior angles**. Alternating interior angles are equal. In the figure above, $\angle d$ and $\angle e$ form a pair of alternating interior angles, and $\angle c$ and $\angle f$ form another pair.

D. Find the measurement of each of the angles below. Use the definitions of the types of angles found in the lesson. The two parallel lines are cut by a transversal. $\angle a = 72°$.

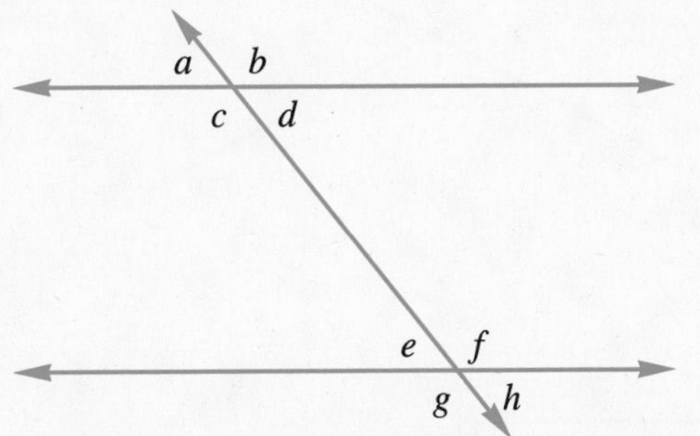

It is known that $\angle a = 72°$. Since $\angle a + \angle b = 180°$ and $\angle a + \angle c = 180°$, $\angle b = 108°$ and $\angle c = 108°$.
$\angle a$ and $\angle d$ are vertical angles; therefore, $\angle d = 72°$. $\angle a$ and $\angle e$ are corresponding angles; therefore, $\angle e = 72°$. Since $\angle e + \angle f = 180°$ and $\angle e + \angle g = 180°$, $\angle f = 108°$ and $\angle g = 108°$.
$\angle e$ and $\angle h$ are vertical angles; therefore, $\angle h = 72°$.

Two lines are perpendicular if they meet at right angles.

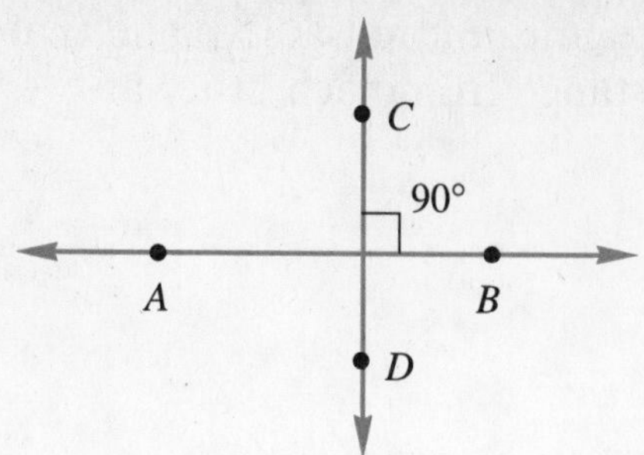

You can write $\overleftrightarrow{AB} \perp \overleftrightarrow{CD}$. This is read "line *AB* is perpendicular to line *CD*." When two parallel lines are cut by a transversal at a right angle, the measurements of all the angles are right angles.

Practice

Solve the following.

1. Are both $\angle 1$ and $\angle 2$ acute angles?

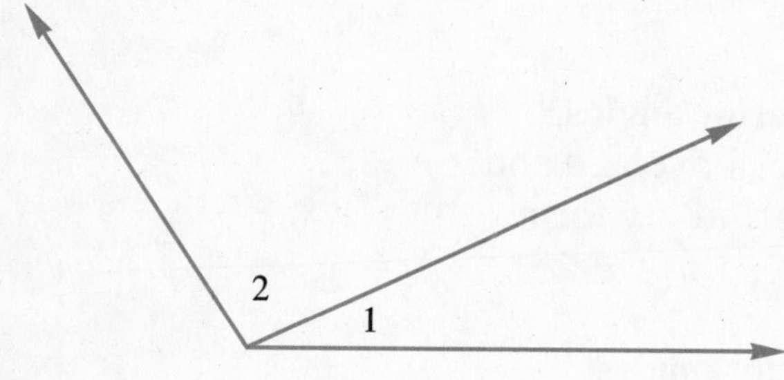

______

2. Are $\angle AOB$ and $\angle AOC$ complementary angles?

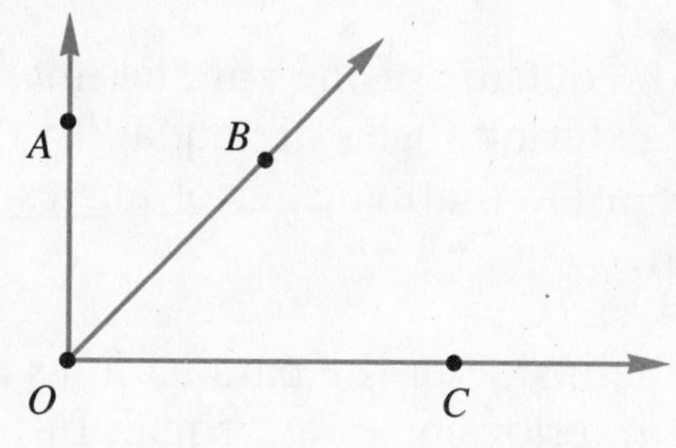

______

3. Name two pairs of supplementary angles.

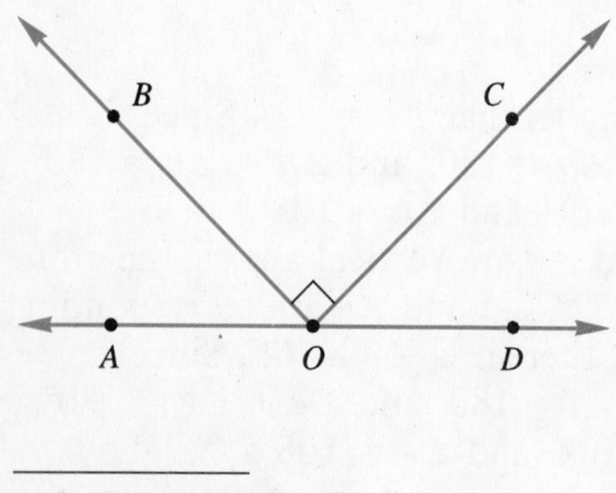

______

______

4. Name two pairs of vertical angles.

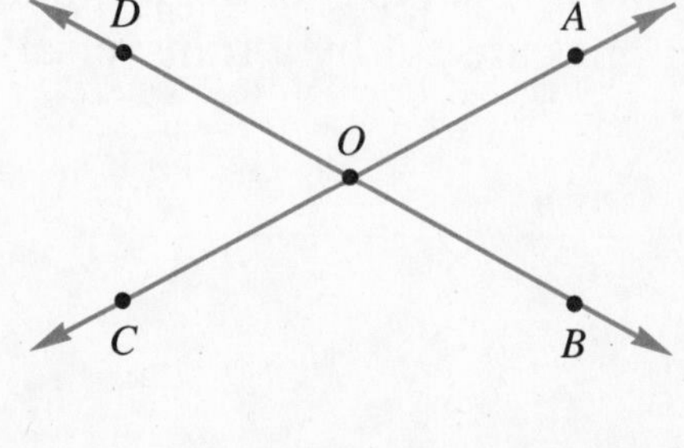

______

______

Use the figure below to answer questions 5–10.

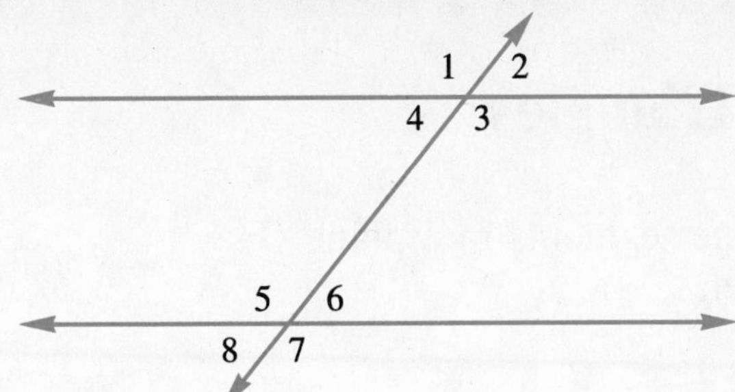

5. Name an acute angle.

______

6. Name an obtuse angle.

______

7. Name two pairs of alternate interior angles.

______

______

8. Name two pairs of alternate exterior angles.

______

______

9. Name two pairs of corresponding angles.

______

______

10. Name two pairs of vertical angles.

______

______

Triangles

A **polygon** is a closed figure formed by line segments in a plane. The sides of a polygon do not cross each other.

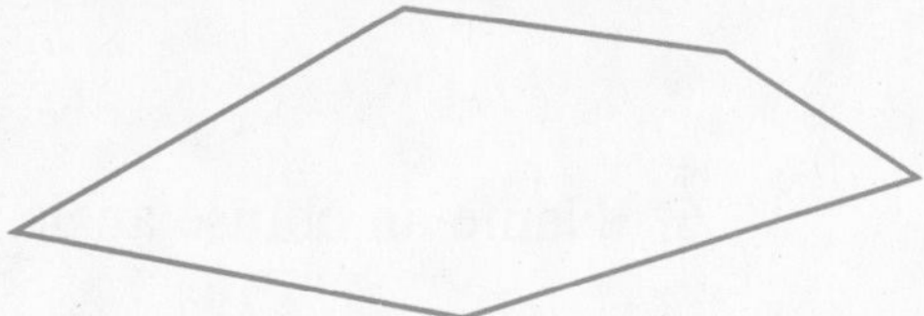

A **triangle** is a polygon with only three sides. Each point where the segments meet is a vertex. The sides are named by using the letters of the vertices at each end of the segment.

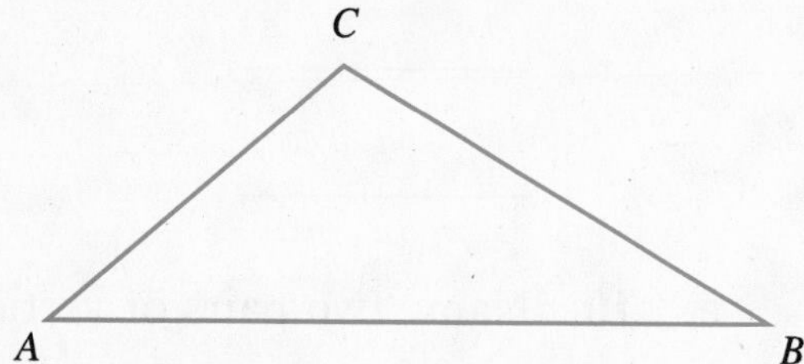

A triangle is named by the letters at the three vertices. Thus $\overline{AB}$, $\overline{BC}$, and $\overline{CA}$ are sides of $\triangle ABC$.

Some triangles have special names. The names refer to special sides or special angles of the triangles.

Triangles With Special Sides

A **scalene triangle** is a triangle with all three sides different in length.

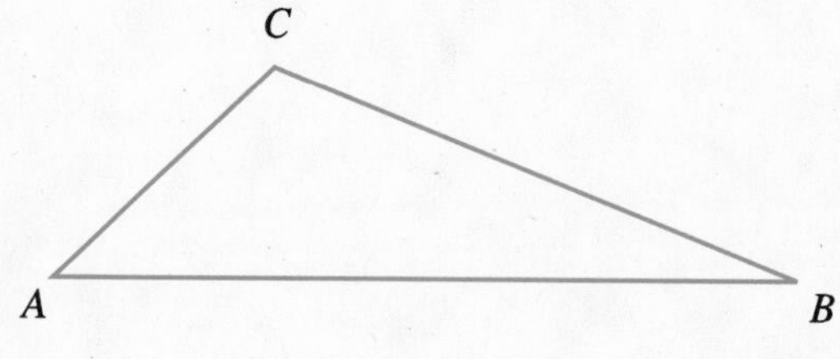

An **isosceles triangle** is a triangle with two equal sides. The angles opposite the two equal sides are called base angles and are equal to each other.

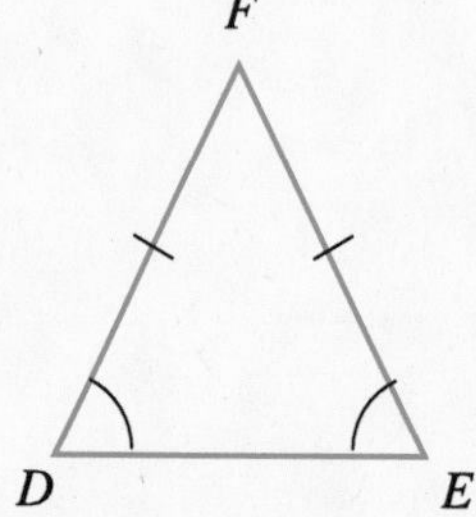

 For more information, see Book 5, pages 55–59.

An **equilateral triangle** is a triangle with all three sides equal. The angles are equal with a measure of 60°.

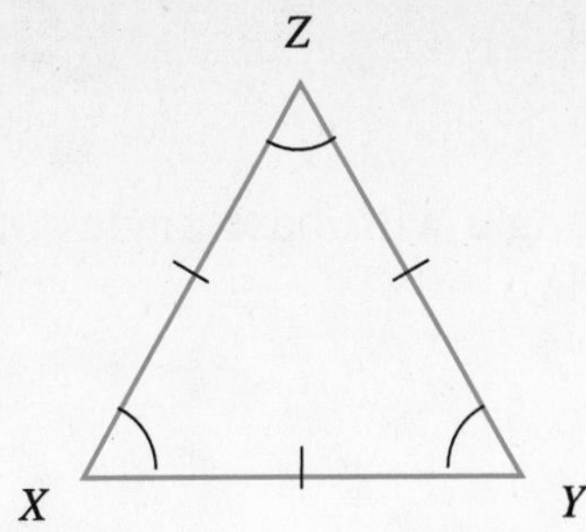

Triangles With Special Angles

An **acute triangle** is a triangle in which each angle measures less than 90°.

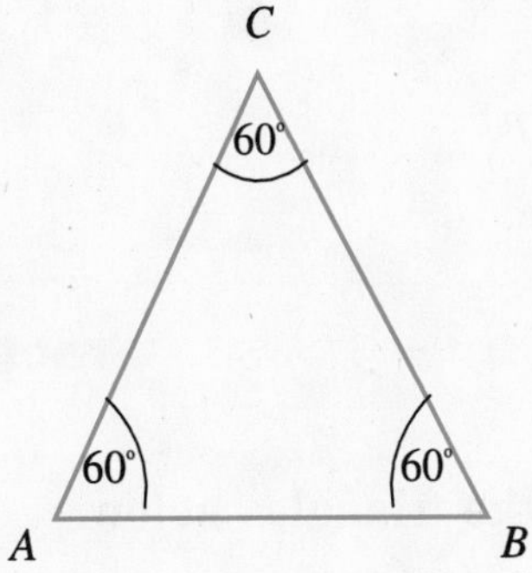

An **obtuse triangle** is a triangle with one angle that measures greater than 90°.

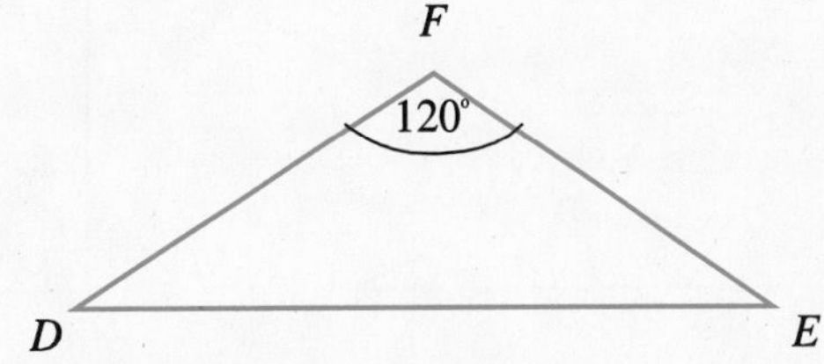

A **right triangle** is a triangle with one right angle.

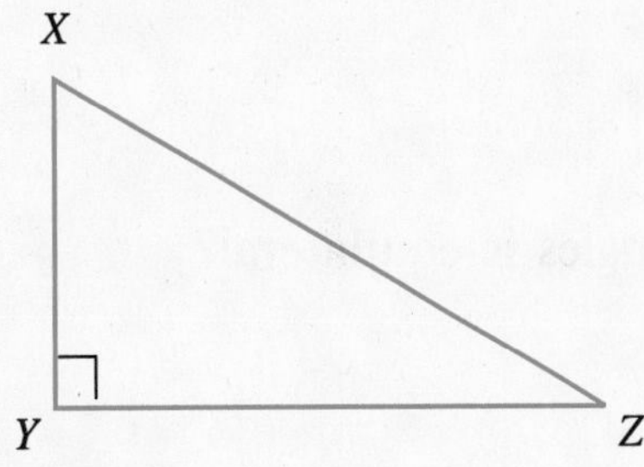

The sum of the three angles in a triangle is always 180°. If you know the measures of two angles of a triangle, you can use this sum to find the measure of the third angle.

Example

$\triangle ABC$ is an isosceles triangle with base angles of 50°. Find the measure of the third angle.

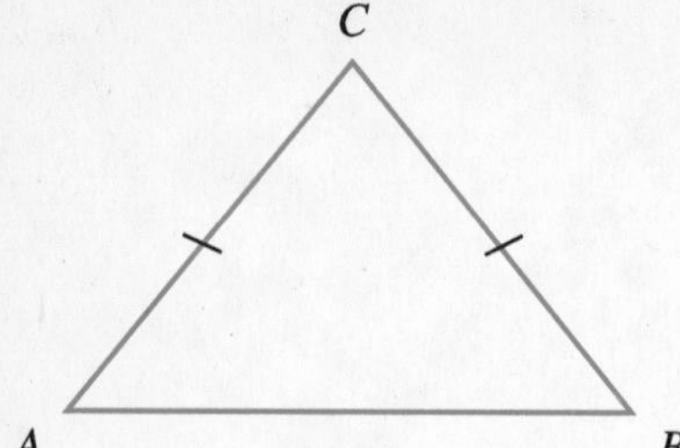

$\angle A = 50°$ and $\angle B = 50°$
$\angle A + \angle B = 100°$
$180° - 100° = 80°$
So, $\angle C = 80°$

Practice

Solve the following using the triangles below.

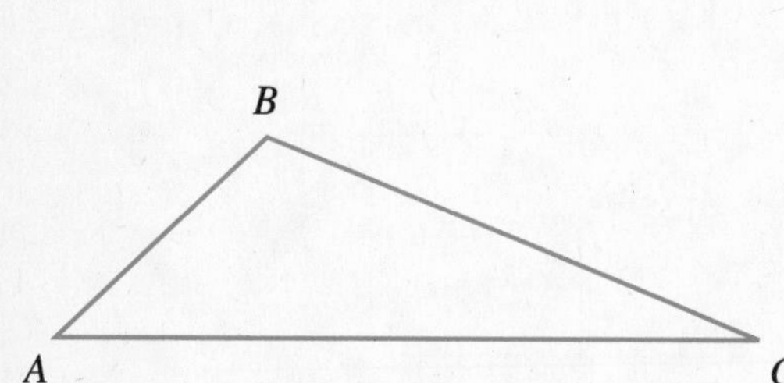

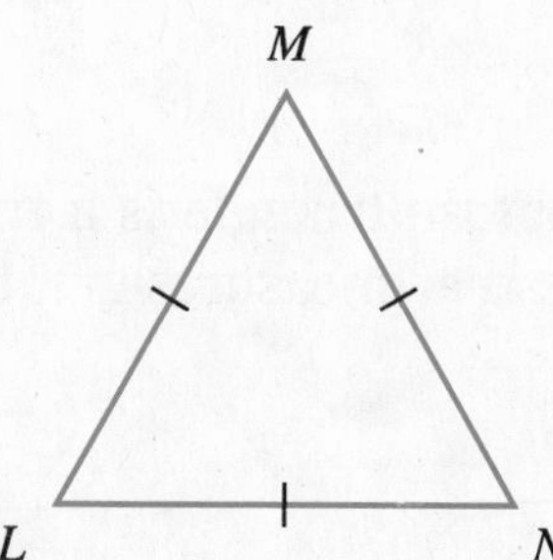

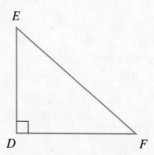

1. Which of the triangles is a right triangle?

____

2. Which of the triangles is an obtuse triangle?

____

3. Which of the triangles is equilateral?

____

LIFE SKILL

Steepness of Hills

The grade of a hill is its steepness. It is important to know the grade of a hill when constructing turns on a road or trails on a ski slope. If a road is too steep, it will be hard for cars to stay on the road. If a trail is too steep, it will be dangerous for a skier.

To determine the percent of grade for a road, do the following:

- ▶ Measure a horizontal distance.
- ▶ Find how much the height has changed over the measured distance.
- ▶ Divide the change in height by the horizontal distance.
- ▶ Multiply by 100 to get the percent of grade.

A distance of 20 feet with a change in height of 1 foot has a percent of grade as follows.

$$\begin{aligned}\text{percent of grade} &= \frac{1}{20} \times 100 \\ &= 0.05 \times 100 \\ &= 5\%\end{aligned}$$

The percent of grade is 5%.
The slope of a hill, or a ski slope, can be calculated the same way.

1. Three ski trails are shown in the figure. Find the percent of grade for each trail.

   Trail A __________

   Trail B __________

   Trail C __________

2. Which slope has the steepest percent of grade?

   ____

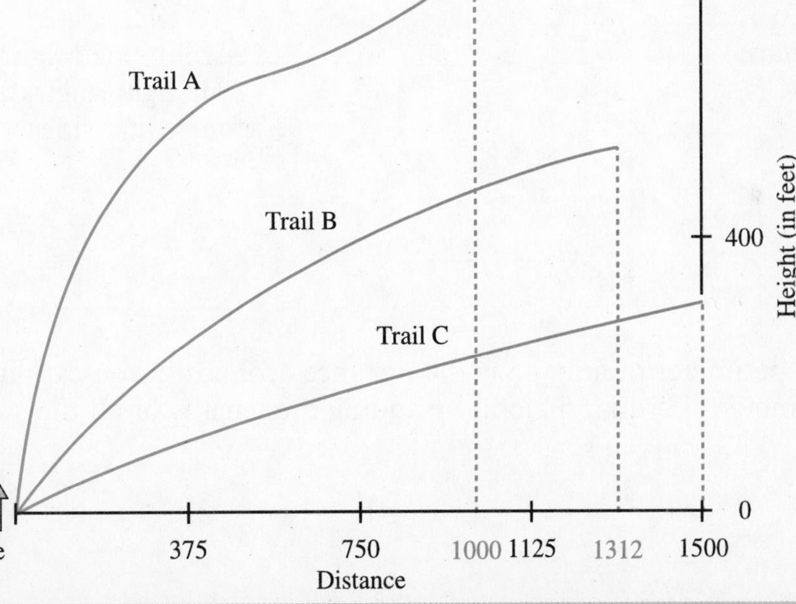

Quadrilaterals and Perimeters

A quadrilateral is a closed figure with four sides. Some quadrilaterals have special names.

Name	Figure	Definition
Trapezoid *ABCD*	D C A B	A four-sided figure with one pair of parallel sides called bases.
Parallelogram *ABCD*	D C A B	A four-sided figure with two pairs of opposite sides that are parallel and congruent.
Rhombus *ABCD*	D C A B	A four-sided figure with four congruent sides and two pairs of opposite sides that are parallel.
Rectangle *ABCD*	D C A B	A four-sided figure with four right angles and two pairs of opposite sides that are equal and parallel.
Square *ABCD*	D C A B	A four-sided figure with four right angles and four equal sides.

The perimeter of a figure is the distance around it. You can find the perimeter of a quadrilateral by adding the lengths of all the sides.

 For more information, see Book 5, pages 60–62.

Example

Find the perimeter of rectangle *ABCD*.

$18 + 20 + 18 + 20 = 76$
Think 2 times the width plus 2 times the length equals the perimeter.

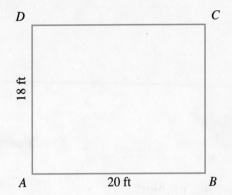

Practice

Find the perimeter for each of the following.

1.

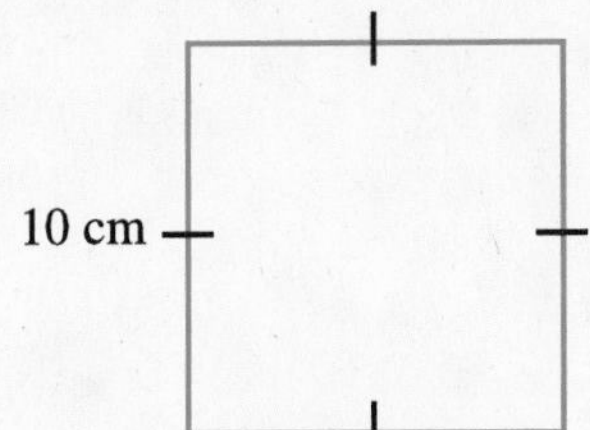

______

2.

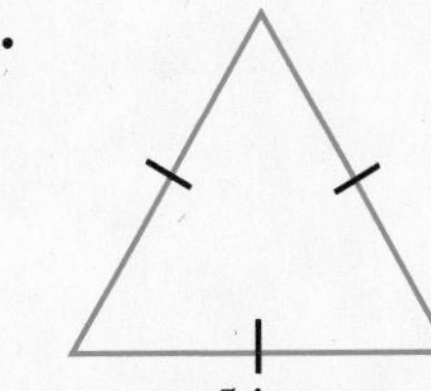

______

3.

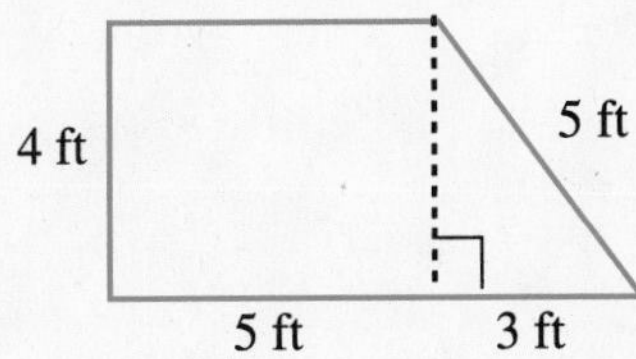

______

4.

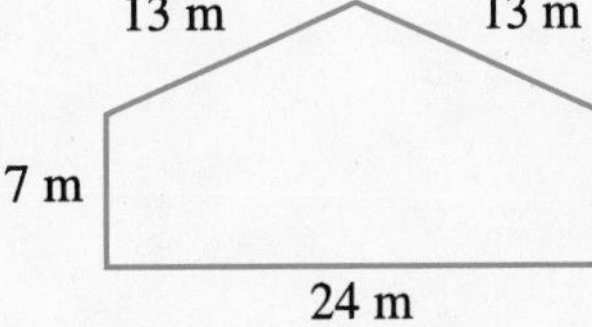

______

5.

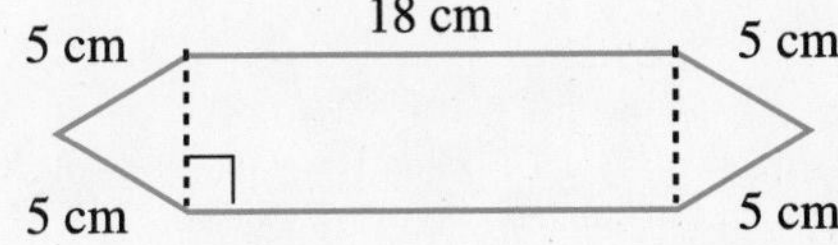

______

6.

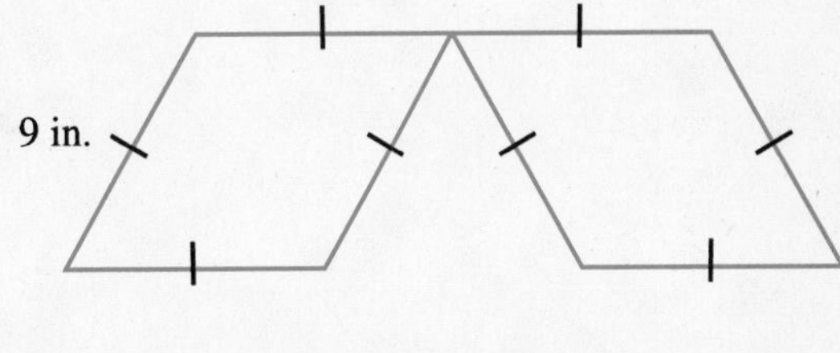

______

7. 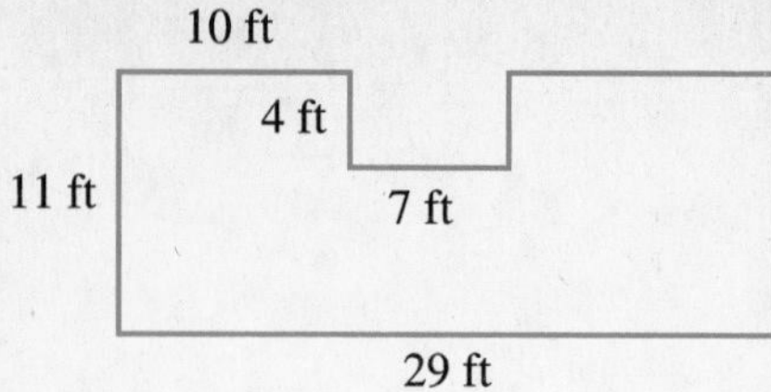

____

8. 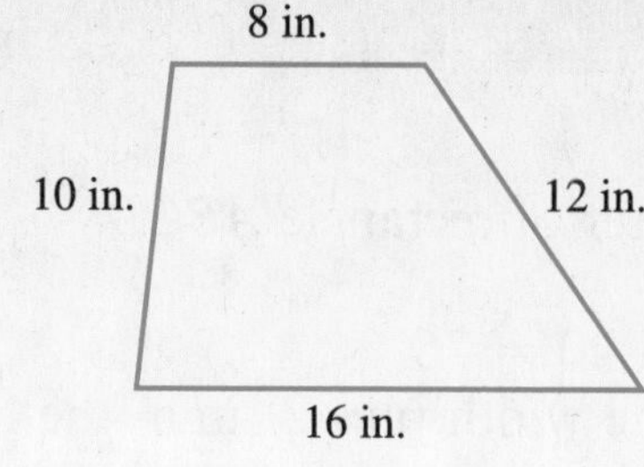

____

Problem Solving

Solve the following problem.

9. A garden has the dimensions as shown in the figure to the right. Find the length of the fencing needed to enclose the garden.

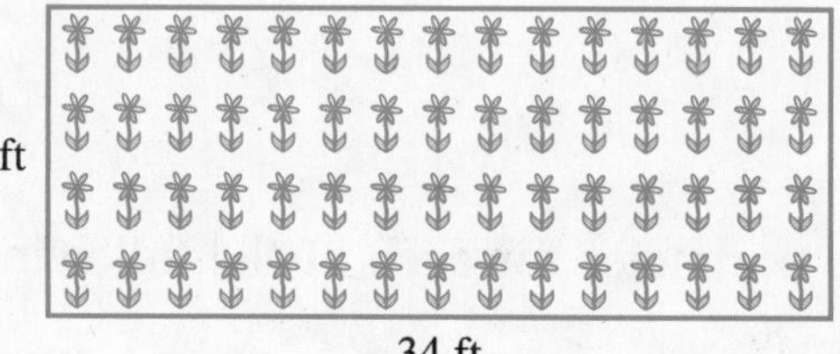

____

Finding Areas

Area is the amount of surface that a figure covers. Area is written using square units such as square feet or square yards.

The formulas for finding the areas of some figures are given at right.

Name of Figure	Formula
Rectangle	$A = lw$ where A is area, l stands for length, and w stands for width
Square	$A = s^2$ where s stands for side
Parallelogram	$A = bh$ where b stands for base and h stands for height
Trapezoid	$A = \frac{1}{2}h(b_1 + b_2)$ where h stands for height and b_1 and b_2 stand for the bases
Triangle	$A = \frac{1}{2}bh$ where b stands for base and h stands for height

Examples

A. Find the area of the figure below.

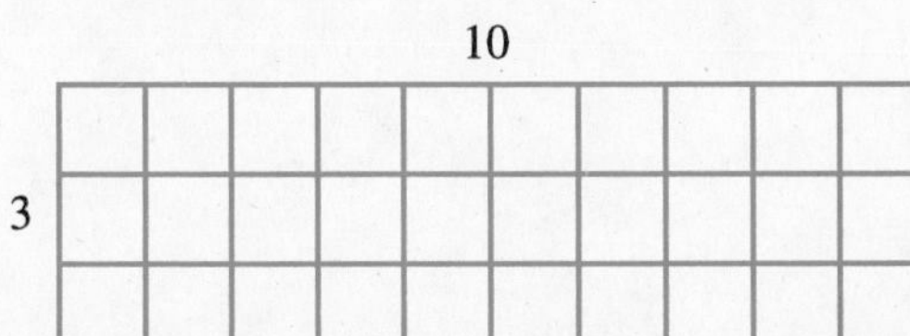

$A = lw$

$A = 3(10)$

$A = 30$

The area is 30 square units.

You can use the formulas above to find the area of a figure that is made up of more than one kind of figure. To do this, draw extra lines to divide the figure into simpler figures. Then find the area of each part. Combine the area of each part to give you the area for the entire figure.

For more information, see Book 5, pages 63–64.

B. Find the area of the figure below.

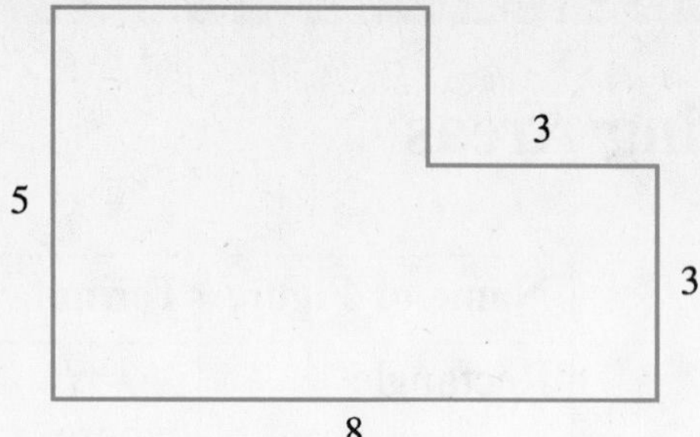

Step 1 Divide the figure into 2 parts.

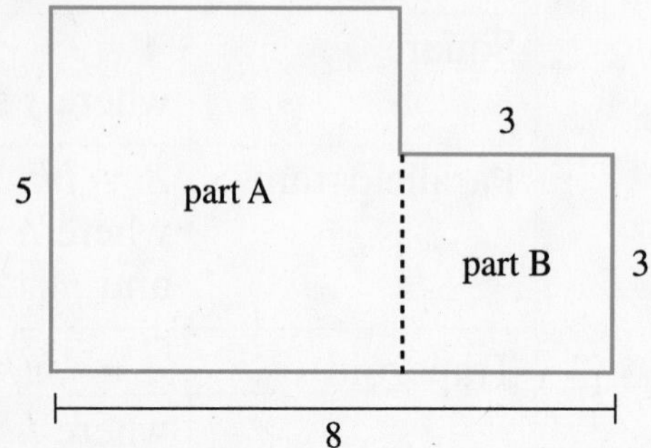

Step 2 Find the area of part A.
The area of part A is 5(5), or 25 square units.

Step 3 Find the area of part B.
The area of part B is 3(3), or 9 square units.

Step 4 Find the total area.
The total area is 25 + 9, or 34 square units.

Practice

Find the area of each figure.

1.

5 cm

18 cm

____

2.

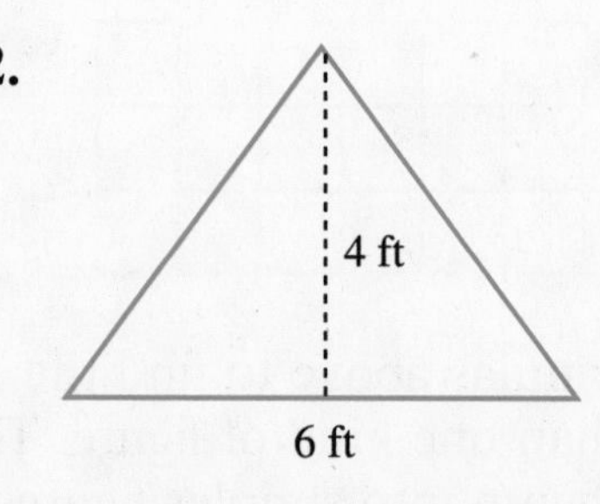

____

3.

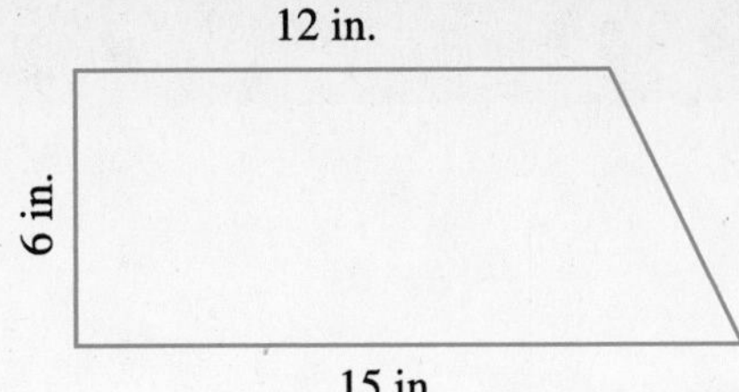

4.

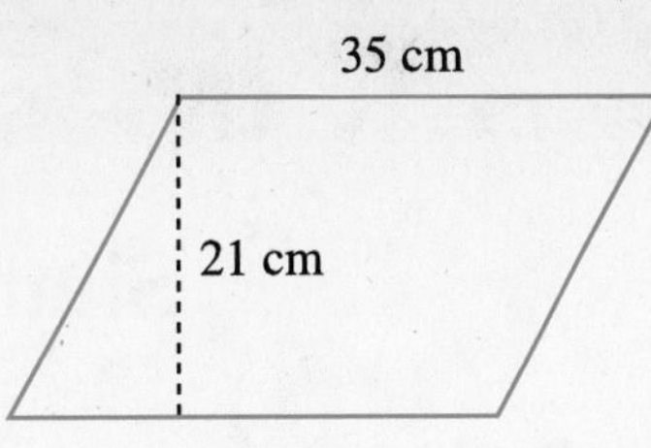

5.

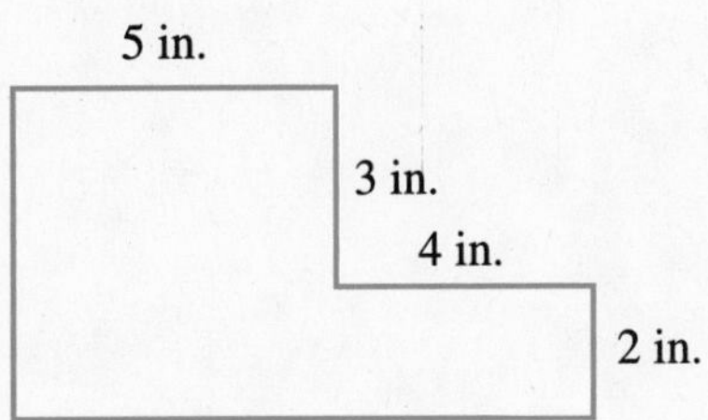

6.

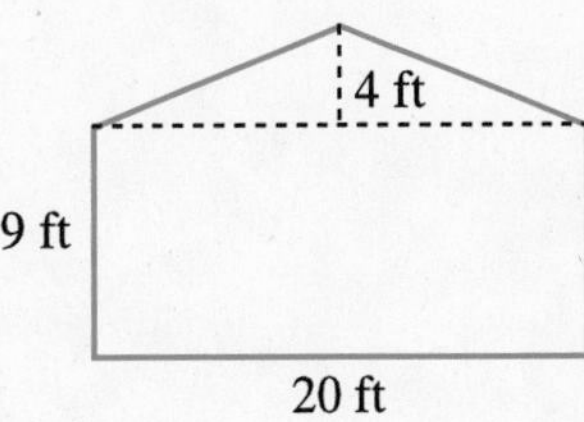

7.

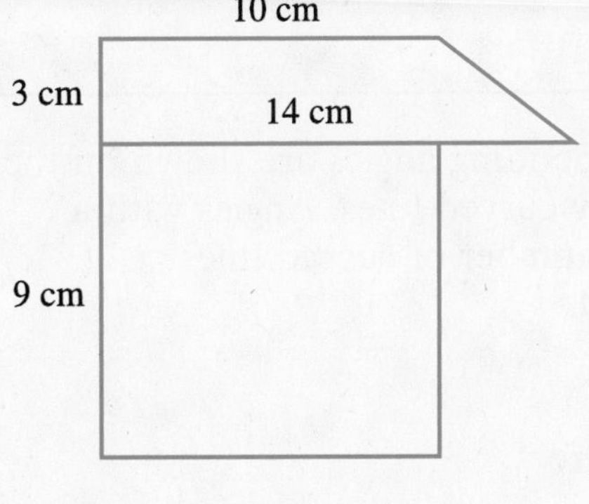

8.

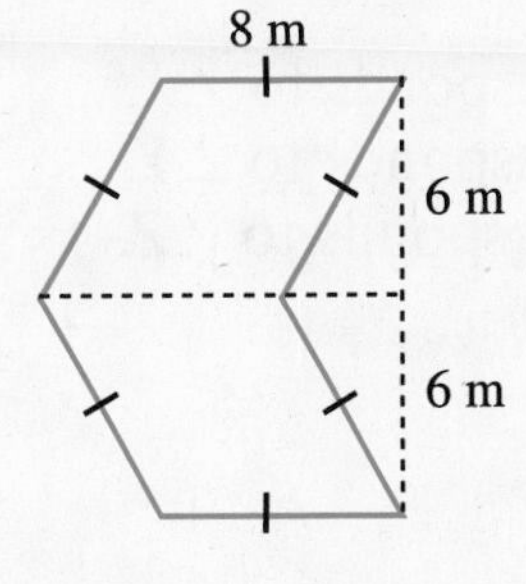

Problem Solving

Solve the following problem.

9. Sylvia wants to repaint one wall in her office. The wall has the measurements shown in the figure. What is the area to be painted?

7 ft

20 ft

LESSON 43

Similar Triangles

Two figures are **similar** if they have the same shape. This means the measures of the angles of one figure are equal to the measures of the angles of the other figure.

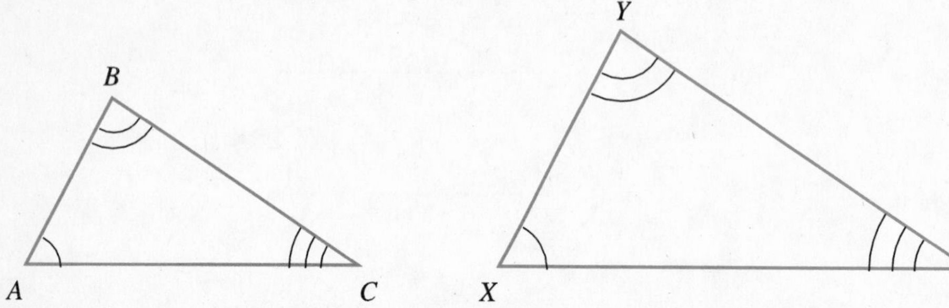

The angles with the same measure are corresponding angles.

$\angle A = \angle X$, so $\angle A$ corresponds to $\angle X$.
$\angle B = \angle Y$, so $\angle B$ corresponds to $\angle Y$.
$\angle C = \angle Z$, so $\angle C$ corresponds to $\angle Z$.

MATH HINT

Corresponding angles are shown on the triangles by curved lines. Angles with a matching number of curved lines correspond.

When referring to similar triangles, corresponding vertices are named in the same order.

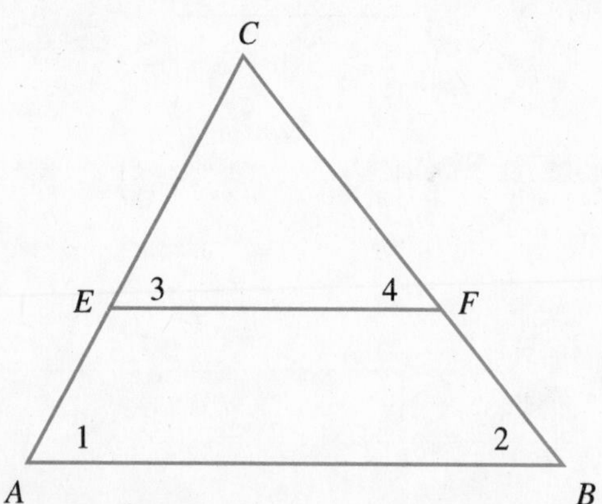

Similar triangles may share parts. In $\triangle ABC$, $\overline{EF}$ is parallel to the base, $\overline{AB}$. The sides AC and BC are transversals passing through the parallel sides.

By corresponding angles, $\angle 1$ is equal to $\angle 3$, $\angle 2$ is equal to $\angle 4$, and $\angle ACB$ is equal to $\angle ECF$. Therefore, $\triangle ABC \sim \triangle EFC$, or $\triangle ABC$ is similar to $\triangle EFC$.

 For more information, see Book 5, pages 67–71.

The sides of similar triangles are in proportion to each other. To find the unknown side measurements of similar triangles, write the proportion and solve for the missing parts.

Examples

A. $\triangle ABC \sim \triangle DEF$. Find the measurements of DE and DF. Let x = the measurement of side DF. Let y = the measurement of side DE.

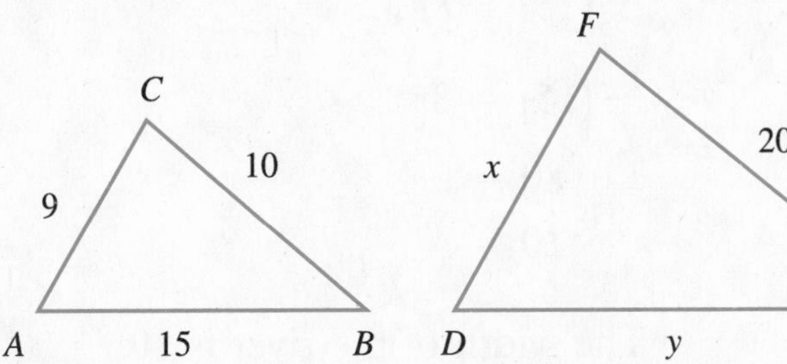

$$\frac{AB}{BC} = \frac{DE}{EF} \qquad \frac{AC}{BC} = \frac{DF}{ED}$$

$$\frac{15}{10} = \frac{y}{20} \qquad \frac{9}{10} = \frac{x}{20}$$

$$30 = y \qquad 18 = x$$

The measurement of side DE is 30.

The measurement of side DF is 18.

When you cannot directly measure a distance, it is often possible to use similar figures and proportions to calculate the measurement needed.

B. Building A is 28 feet high and casts a shadow of 15 feet. Next to Building A is Building B, which casts a 60-foot shadow. What is the height of Building B?

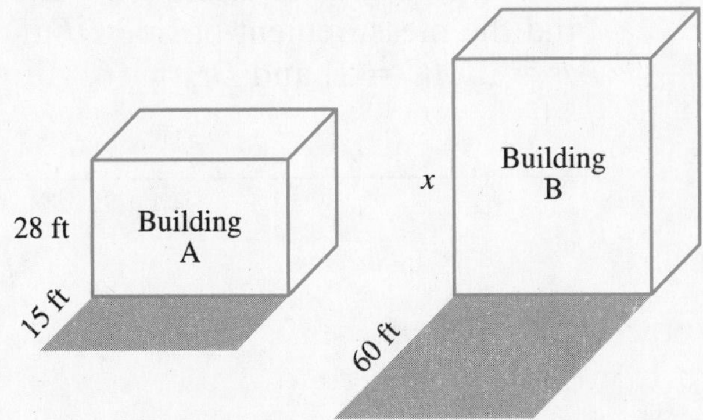

Set up a proportion.

$$\frac{\text{height of Building A}}{\text{shadow of Building A}} = \frac{\text{height of Building B}}{\text{shadow of Building B}}$$

$$\frac{28}{15} = \frac{x}{60}$$

$$x = \frac{28 \times 60}{15}$$

$$x = 112$$

The height of Building B is 112 ft.

C. A scout troop estimated the distance across a river by making the measurements shown below. How wide is the river?

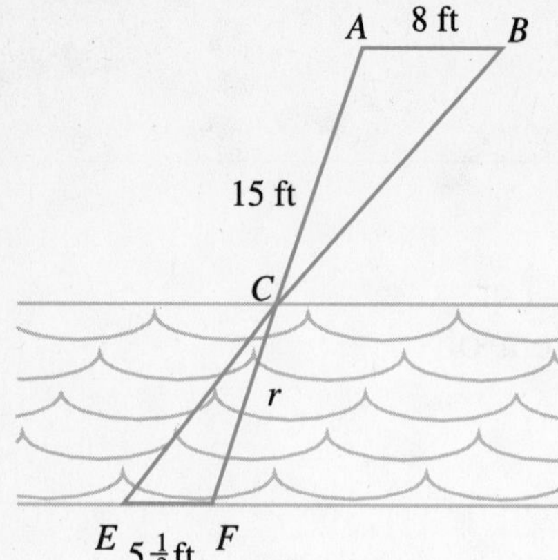

Write a proportion using corresponding sides. Then substitute the known measurements. Let r = the distance across the river. (Hint: $CF = r$)

$$\frac{FE}{AB} = \frac{CF}{AC}$$

$$\frac{5\frac{1}{3}}{8} = \frac{r}{15}$$

$$5\tfrac{1}{3}(15) = 8r$$

$$\frac{16}{\cancel{3}_1}(\cancel{15}^5) = 8r$$

$$80 = 8r$$

$$10 = r$$

The width of the river is 10.

Practice

Solve the following.

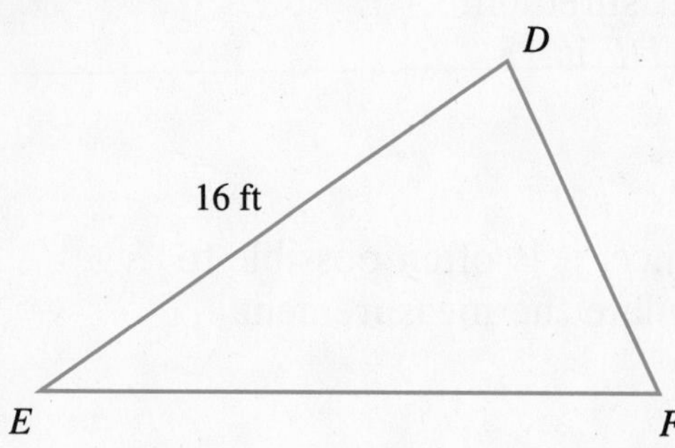

$\Delta DEF \sim \Delta ABC.$

1. For the triangles DEF and ABC, name the corresponding angles and sides.

____

2. Refer to triangles ABC and DEF above. Find the measurement of side DF if $AB = 2$, $AC = 1$, and $DE = 16$.

____

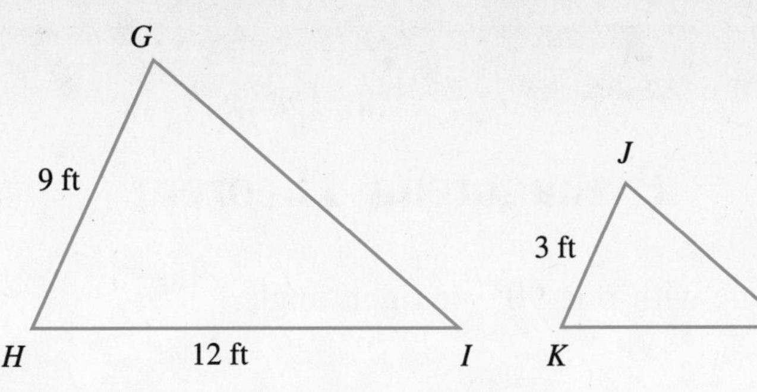

$\triangle GHI \sim \triangle JKL$.

3. For the triangles *GHI* and *JKL*, name the corresponding angles and sides.

4. Refer to triangles *GHI* and *JKL* above. Find the measurement of side *KL* if $GH = 9$, $HI = 12$, and $JK = 3$.

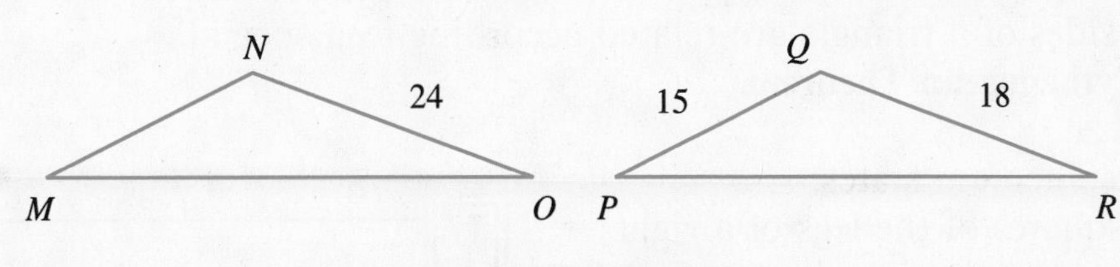

$\triangle MNO \sim \triangle PQR$.

5. For the triangles *MNO* and *PQR*, name the corresponding angles and sides.

6. Refer to triangles *MNO* and *PQR* above. Find the measurement of side *MN* if $NO = 24$, $PQ = 15$, $RQ = 18$.

Problem Solving

Solve the following problems.

7. A photograph measures 40 mm by 30 mm. It is enlarged so that the longer side measures 120 mm. What is the measurement of the shorter side?

8. A painting that measures 36 inches by 45 inches is reproduced with a width of 20 inches. Find the length of the reproduction.

Pythagorean Theorem

A **right triangle** is a triangle with one 90°, or right, angle.

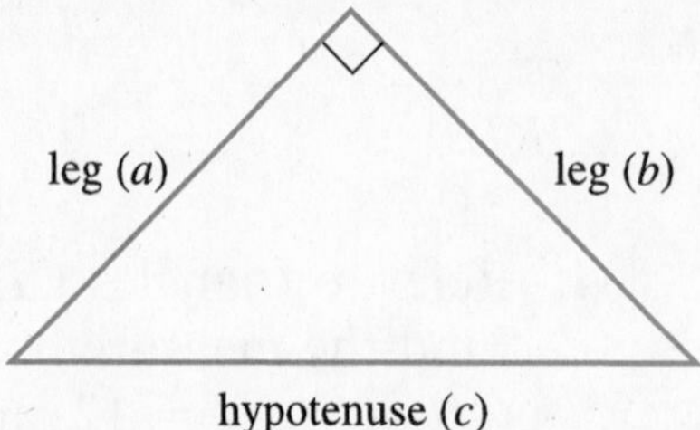

The two shorter sides of a right triangle are called **legs**. The longest side of a right triangle is called the **hypotenuse** and is opposite the right angle. The sides of a triangle are related according to a special rule called the **Pythagorean Theorem**.

The Pythagorean Theorem states:
The sum of the squares of the legs of a right triangle equals the square of the hypotenuse.

$$a^2 + b^2 = c^2$$

You can use the Pythagorean Theorem to find the measures of the sides of a right triangle.

MATH HINT

You can use either a table or a calculator to find the square or square root of a number. The x^2 key is used to find the square of a number. The $\sqrt{x}$ key is used to find the square root of a number.

Examples

A. In the right triangle below, the hypotenuse is 20 inches and one of the legs is 12 inches. Find the measurement of the missing leg. (Hint: Let b = the measurement of the missing leg.)

$$a^2 + b^2 = c^2$$

Since $a = 12$ and $c = 20$,

$$(12)^2 + b^2 = (20)^2$$
$$144 + b^2 = 400$$
$$b^2 = 400 - 144$$
$$b^2 = 256$$
$$b = 16$$

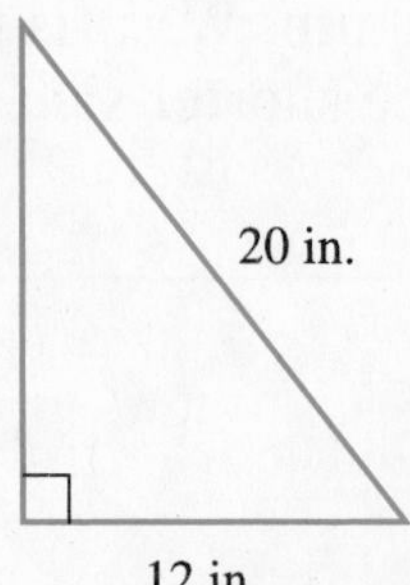

 For more information, see Book 5, pages 72–74.

B. A right triangle has legs of 16 feet and 22 feet. Find the measurement of the hypotenuse. (Hint: Let c = the measurement of the hypotenuse.)

$$a^2 + b^2 = c^2$$
$$(16)^2 + (22)^2 = c^2$$
$$256 + 484 = c^2$$
$$740 = c^2$$
$$27.20 = c$$

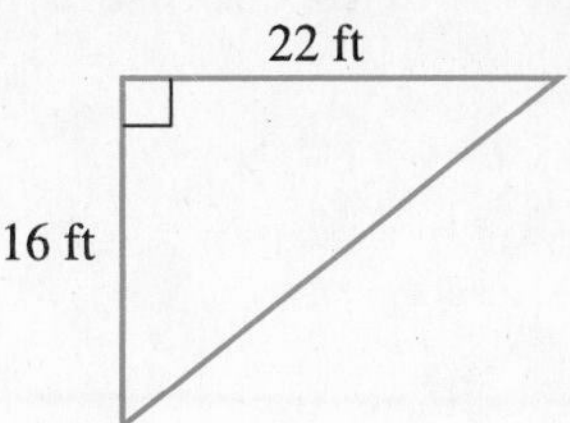

The hypotenuse measures 27.20 feet.

Pythagorean Triples are sets of three whole numbers that fit neatly into the Pythagorean Theorem. Multiples of these triples are also Pythagorean Triples. Examples of these triples are:

3, 4, 5	5, 12, 13	8, 15, 17
7, 24, 25	20, 21, 29	12, 35, 37

C. Use Pythagorean Triples to determine if a triangle with sides measuring 9 cm, 12 cm, and 15 cm is a right triangle.

3, 4, 5 is a Pythagorean Triple.
9, 12, 15 is a multiple of the triple.
$(9 = 3 \times 3;\ 12 = 4 \times 3;\ 15 = 5 \times 3)$

This is a right triangle.

Since rectangles and squares have right angles, two right triangles are formed when you draw one of the diagonals for these figures. The diagonals are the hypotenuse of each right triangle formed. To find the length of the diagonals of these figures, use the Pythagorean Theorem.

D. A rectangle has an 8-inch width and a 15-inch length. What is the measure of the diagonal?

$$a^2 + b^2 = c^2$$
$$(8)^2 + (15)^2 = c^2$$
$$64 + 225 = c^2$$
$$289 = c^2$$
$$17 = c$$

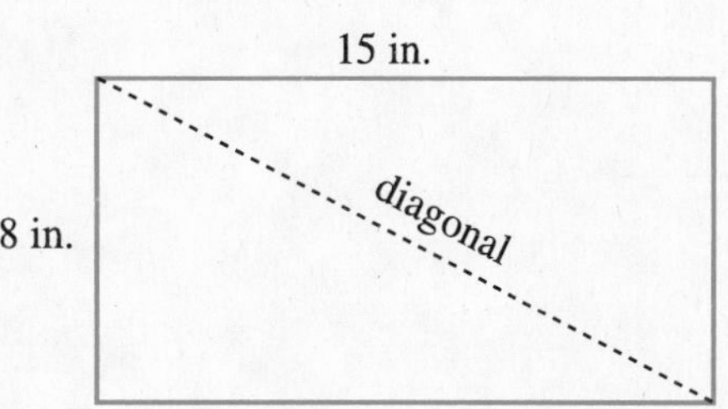

The diagonal measures 17 inches.

Practice

Use the Pythagorean Theorem to complete the table below.

1.

a	b	a^2	b^2	c^2	c
3	4				
6	8				
5	12				
12					13
	21				29

Find the length of the diagonal for each of the figures.

2. 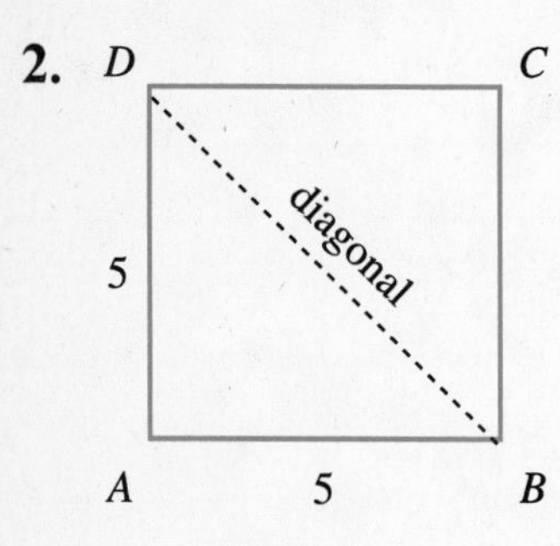

____

3. 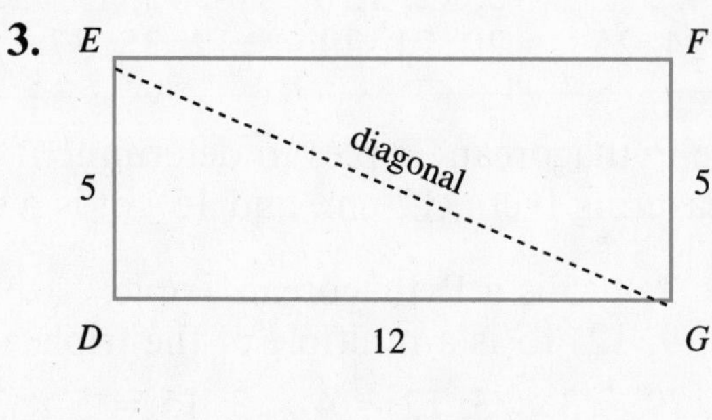

____

4. 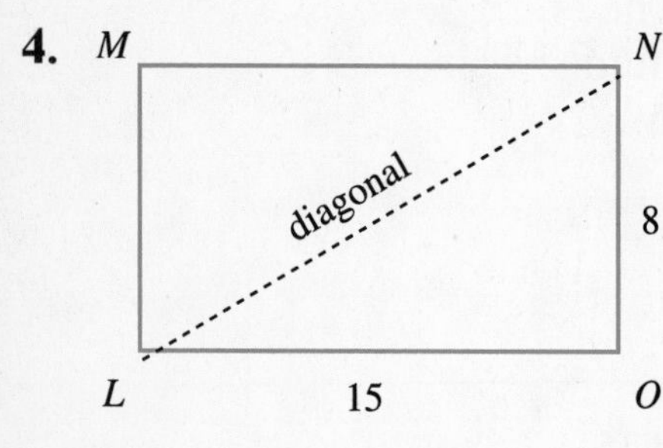

____

5. 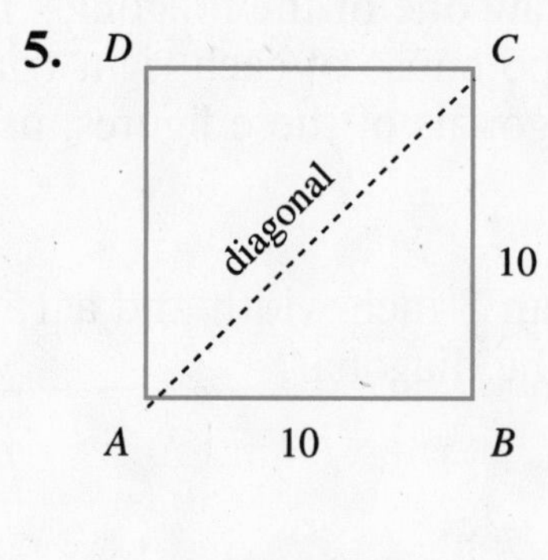

____

Determine if the following triangles are right triangles. Answer yes or no.

6.

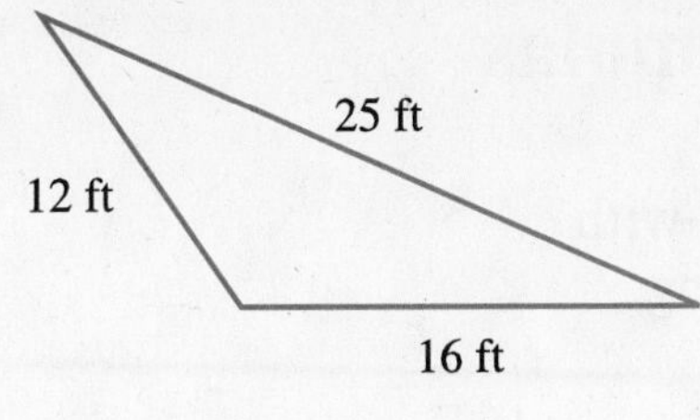

____

7.

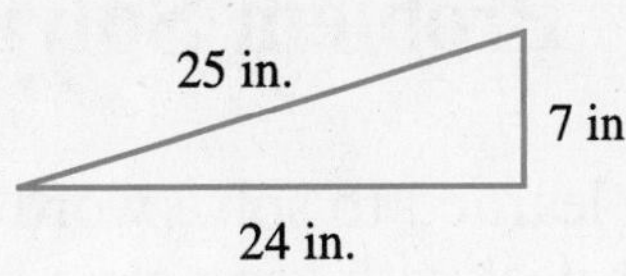

____

8.

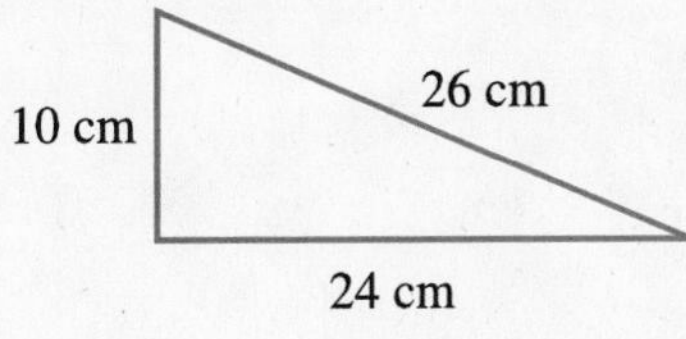

____

9.

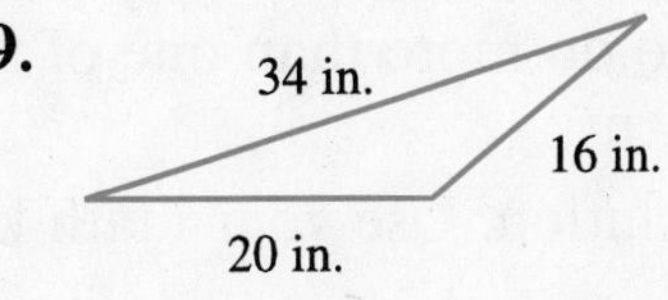

____

Problem Solving—Using Formulas

The steps you have learned to solve word problems can be used with word problems that deal with geometric figures. Use the following steps:

Step 1 Read the problem and underline the key words. These words will usually relate to some mathematical reasoning.

Step 2 Make a plan to solve the problem. Ask yourself, Should I add, subtract, multiply, divide, round, or compare? You may have to do more than one of these operations for the same problem.

Step 3 Find the solution. Use your math knowledge to find your answer.

Step 4 Check your answer. Ask yourself, Is the answer reasonable? Did you find what you were asked for?

Formulas are useful for solving some geometry problems. When you use geometric formulas to solve a problem, first state what each letter represents. This helps you keep the values clear. Then replace the letters with the values you know.

Example

A television antenna has a guy wire that is attached to the top of the antenna and to the eave of the roof, forming a right triangle. The guy wire is 50 feet long and is attached to a location 30 feet from the bottom of the antenna. How tall is the antenna?

Step 1 Read the problem and identify the key words.
The key words are **right triangle**.

Step 2 Make a plan to solve the problem.
Identify the sides of the right triangle and use the Pythagorean Theorem to solve.

Step 3 Find the solution.
The guy wire is the hypotenuse. The antenna is one of the legs of the triangle. (Hint: Let b = the height of the antenna.)

 For more information, see Book 5, pages 190–193.

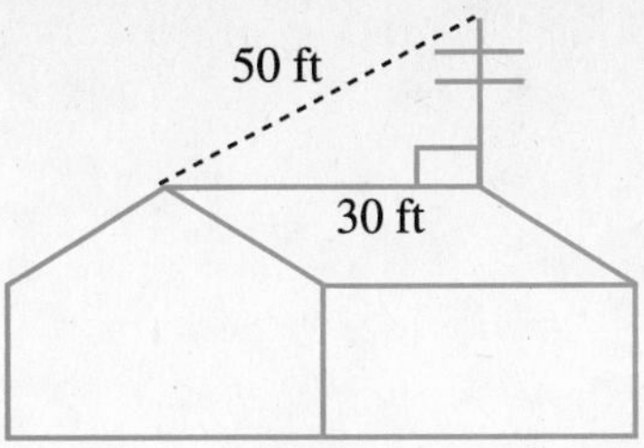

$$c^2 = a^2 + b^2$$
$$(50)^2 = (30)^2 + b^2$$
$$2{,}500 = 900 + b^2$$
$$1{,}600 = b^2$$
$$40 = b$$

The antenna is 40 feet tall.

Step 4 Check your answer.

$$(50)^2 = (30)^2 + (40)^2$$
$$2500 = 900 + 1600$$
$$2500 = 2500$$

The answer checks.

Practice

Solve the following.

1. A square garden is enclosed with 92 feet of fencing. What is the length of each side?

____

x

2. The second angle of a triangle is 20 degrees more than the first angle. The third angle is twice the size of the first angle. If the first angle measures 40 degrees, what are the measures of the other angles? (Hint: The sum of the three angles in a triangle is 180°.)

$x + 20$ $2x$ x

____

____

3. John's house needs a new roof. The roof is made up of two rectangular sides with lengths of 40 feet and widths of 12 feet. What is the area of the roof?

12 ft   40 ft

____

UNIT

6

Posttest

Identify each of the following figures.

1.

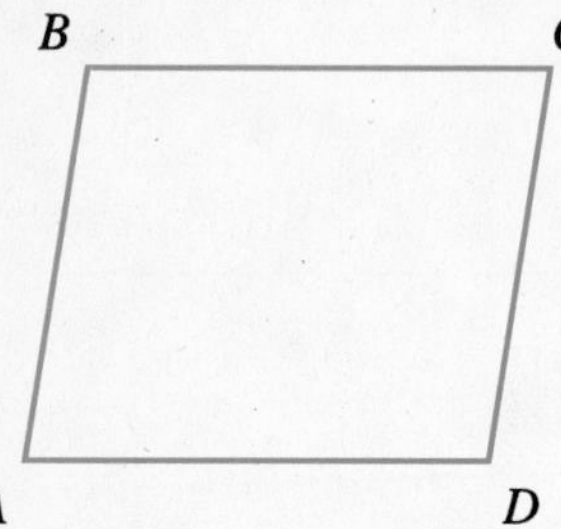

____

2.

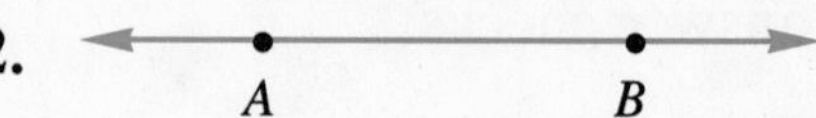

____

3.

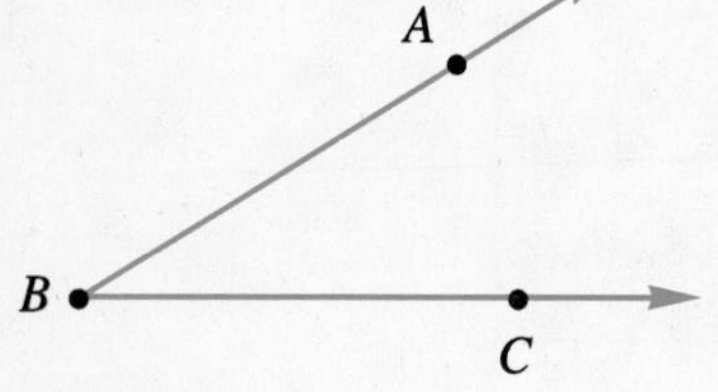

____

4.

____

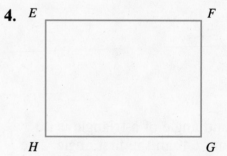

Use the figure below to answer questions 5–8.

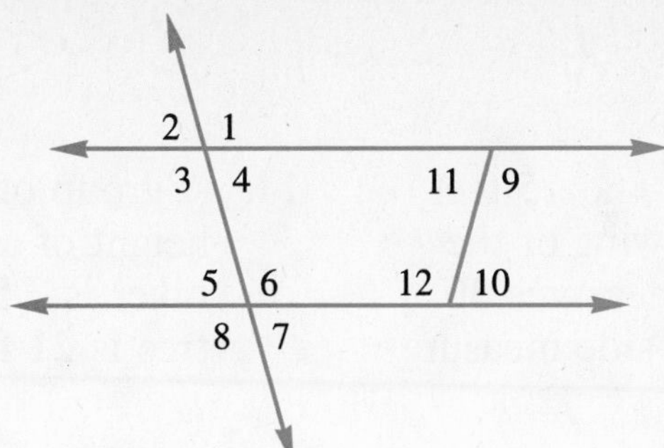

5. Name an acute angle.

____

6. Name an obtuse angle.

____

7. Name two pairs of alternating interior angles.

____

____

8. Name two pairs of alternating exterior angles.

____

____

Describe the sides in each of the following triangles.

9. isosceles triangle _______________

____

10. scalene triangle _______________

____

Find the perimeter and area for each of the following figures.

11.

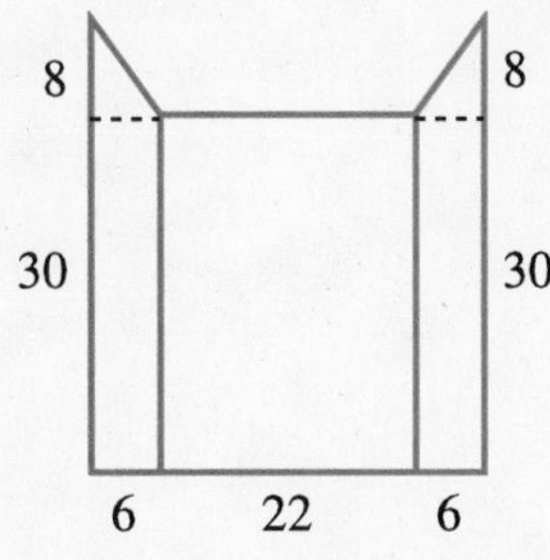

____

____

12.

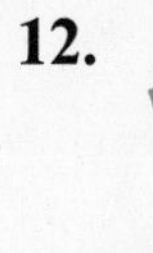

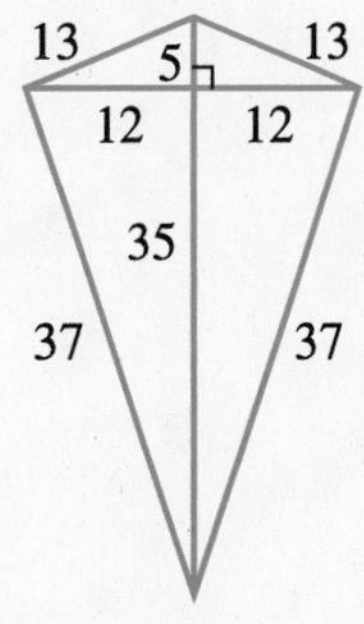

____

____

Solve the following problems.

13. A building is 10 feet wide and 15 feet long at the base. On a drawing of the building, the longer side measures 30 cm. What will the shorter side measure on the drawing?

____

14. A group of hikers need to measure the height of a tree. The shadow of a 5-foot hiker is 3.5 feet. The shadow of the tree is 21 feet. How tall is the tree?

____

Find the diagonals for each of the following.

15. *B* *C*

diagonal

30

A 40 *D*

____

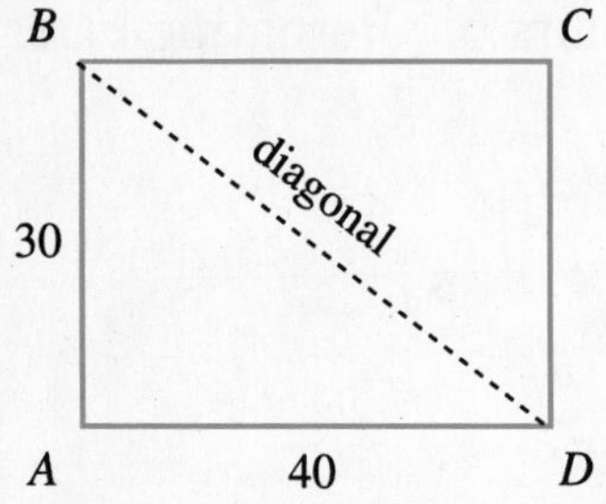

16. *N* *M*

diagonal

8

K 8 *L*

____

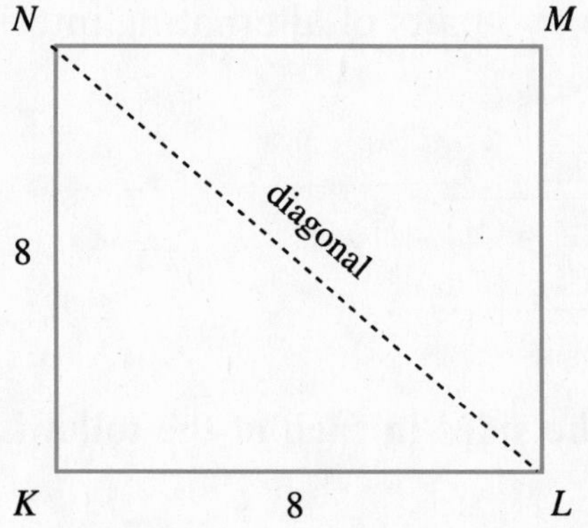

UNIT

Circles

Pretest

Identify the following parts of the circle shown below.

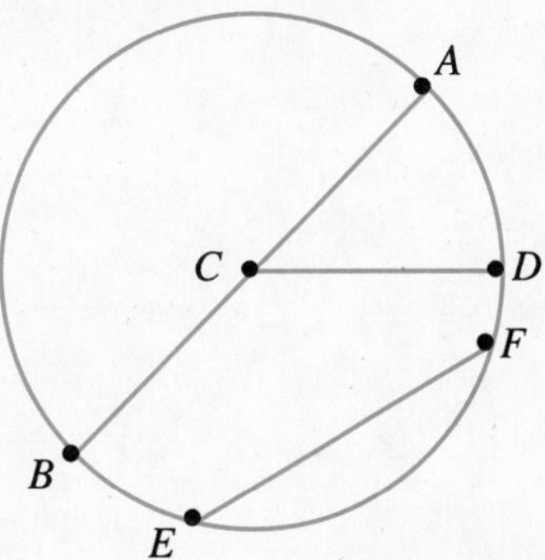

1. radius

   ____

2. diameter

   ____

3. center

   ____

4. chord

   ____

Find the circumference and area for each circle. Use $\pi = 3.14$.

5.

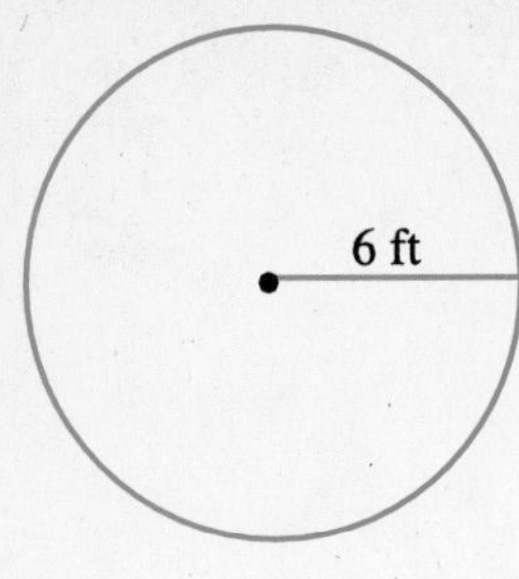

____

____

6.

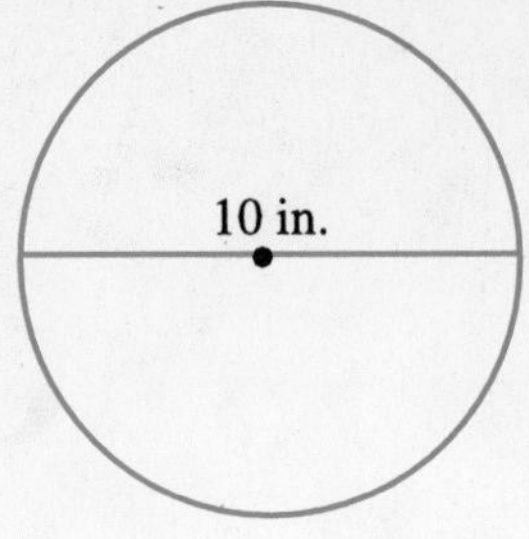

____

____

7.

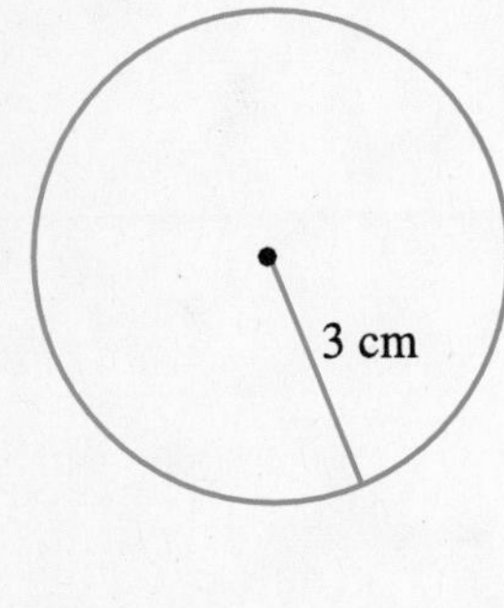

____

____

8.

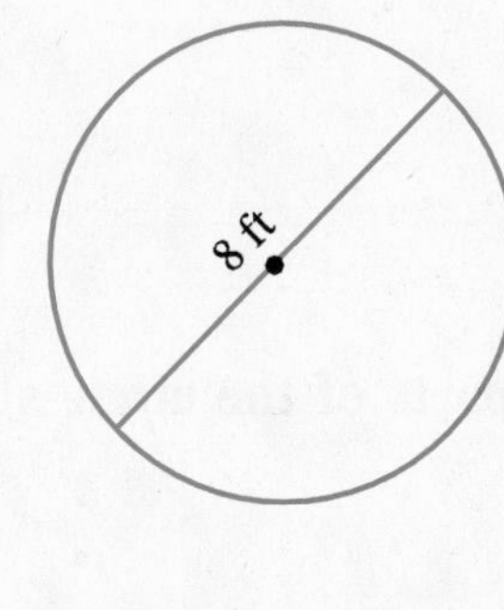

____

____

Problem Solving

Solve the following problems.

9. Sylvia enclosed a circular garden with fencing. The garden has a radius of 5 feet. How much fencing did Sylvia use?

____

10. A CD is 4 inches in diameter. What is its area?

____

LESSON 46

Circles—Parts and Circumferences

A **circle** is a closed figure that includes all points the same distance from a given point, the center.

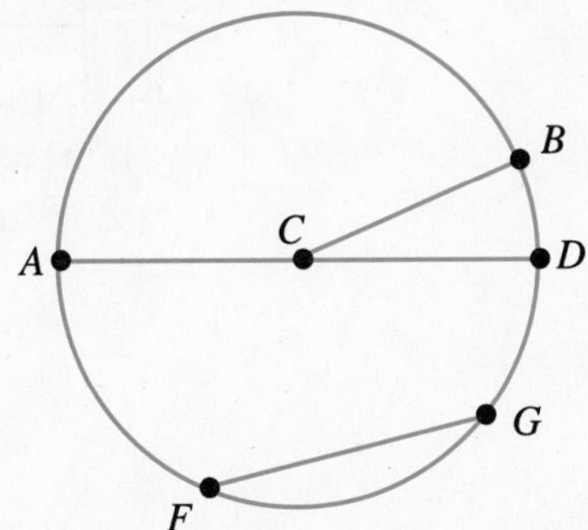

The **diameter**, $\overline{AD}$, is a line segment that passes through the center of the circle and connects two points on the circle.

The **radius**, $\overline{CB}$, is a line segment that connects the center of the circle to a point on the circle. The radius is one-half the length of the diameter, or $r = \frac{1}{2}d$.

Any line segment from one point on a circle to another is a **chord** ($\overline{FG}$).

The distance around the circle is its **circumference.** It is like the perimeter of a polygon. To find the circumference, multiply the diameter by the value of π, which is $\frac{22}{7}$, or about 3.14. So, $C = \pi d$, where C is the circumference and d is the diameter.

Examples

A. The diameter of a circle is 24 inches. Find the radius and the circumference of the circle.

The radius is one-half the diameter.

$$r = \frac{1}{2}d$$
$$r = \frac{1}{2}(24)$$
$$r = 12$$

The radius is 12 inches.

For more information, see Book 5, pages 120–123.

The circumference is the product of π and the diameter.

$$C = \pi d$$
$$C = 3.14(24)$$
$$C = 75.36$$

The circumference is 75.36 inches.

B. Find the circumference of a circle with a diameter of 21 inches.

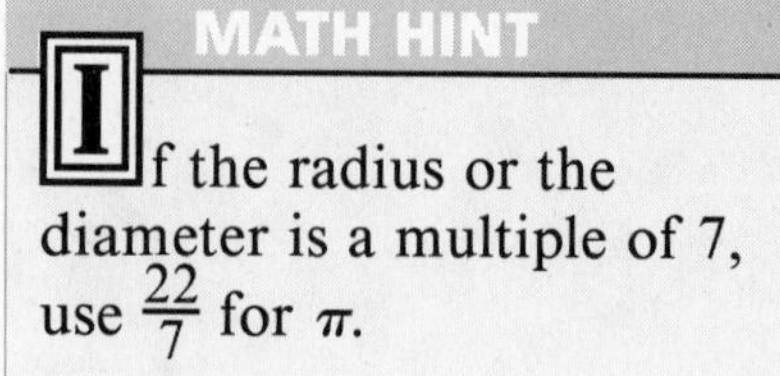
MATH HINT

If the radius or the diameter is a multiple of 7, use $\frac{22}{7}$ for π.

$$C = \pi d$$
$$C = \frac{22}{7} \times \frac{21}{1}$$
$$C = 66$$

The circumference is 66 inches.

Practice

Solve the following.

1. Complete the following table.

Radius	Diameter	Circumference	$\frac{\text{Circumference}}{\text{Diameter}}$
	5		
2			
		31.4	
		6.28	

Find the circumference for each of the following.

2.

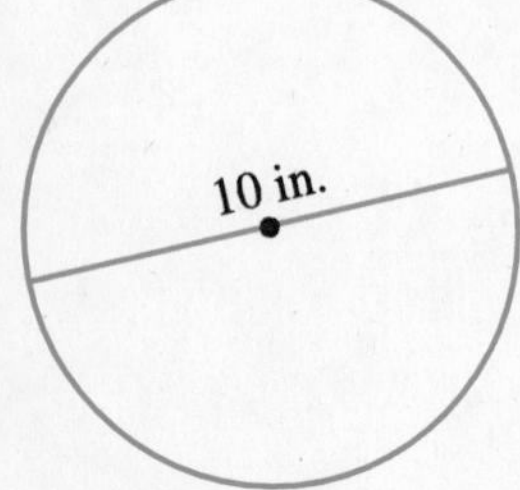

______

3.

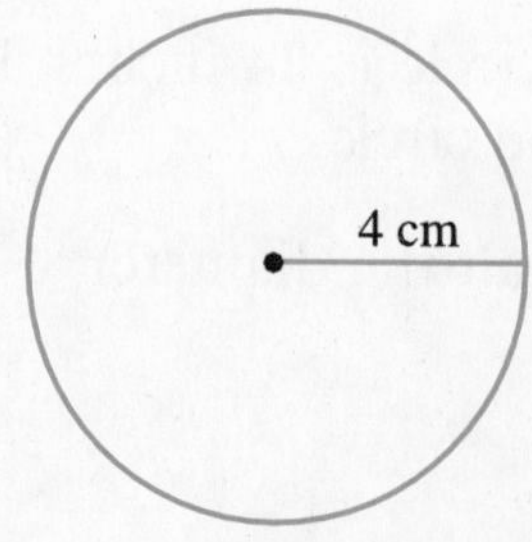

______

4.

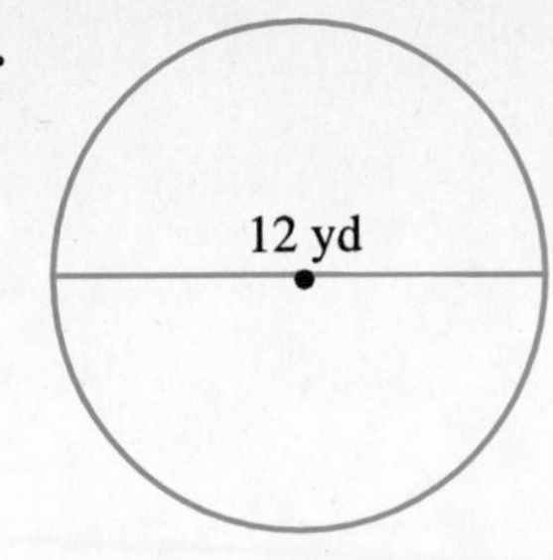

____

5.

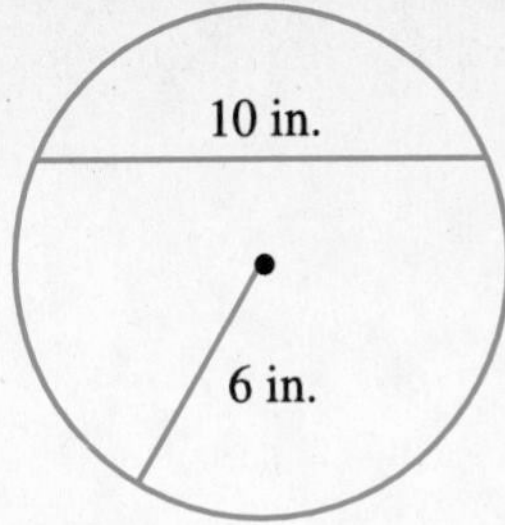

____

6.

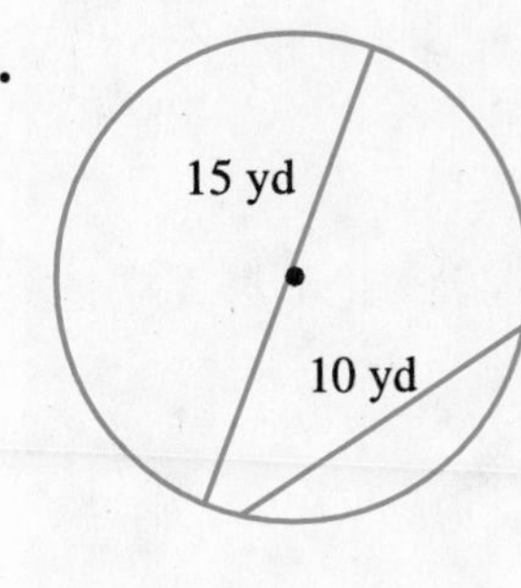

____

7.

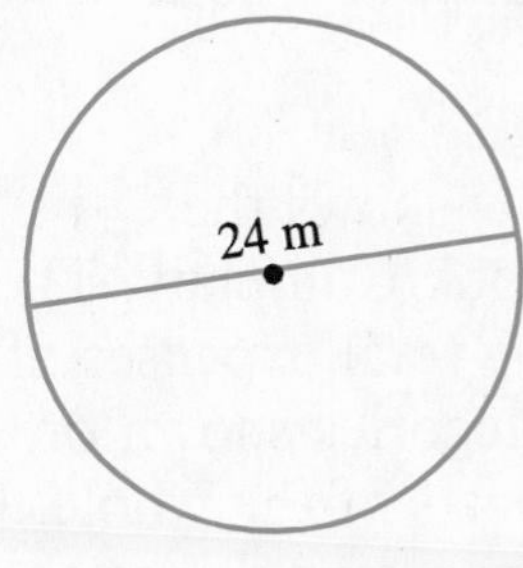

____

Problem Solving

Solve the following problems.

8. A circular fish pond has a 4-foot diameter. The pond is surrounded by a brick border that is 18 inches wide. What is the circumference of the border?

____

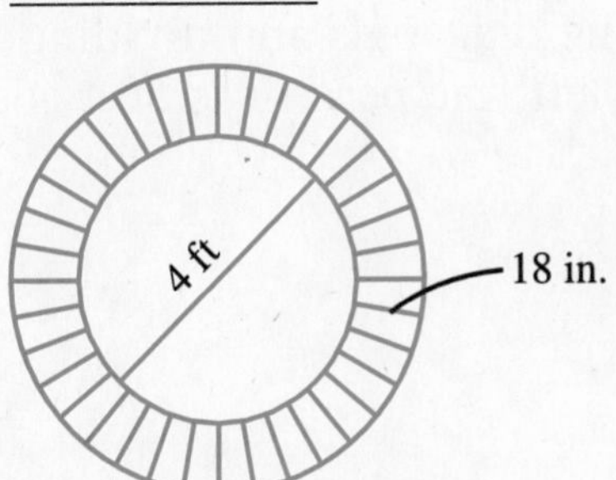

9. Tetherball is played with a ball that is attached to the top of a pole by a rope. The rope is 6 feet long. When the ball is hit, what is the greatest distance the ball can travel in one full swing?

____

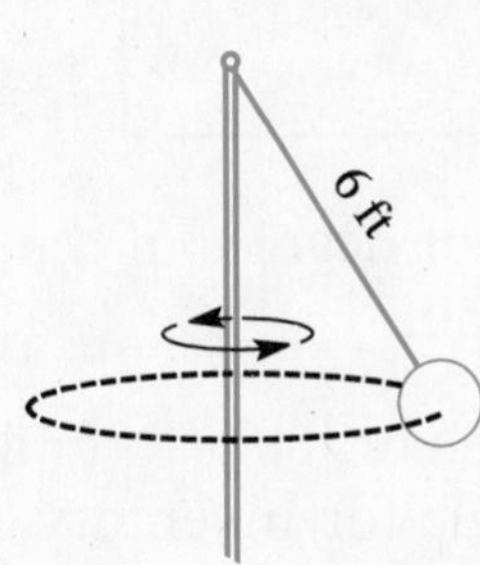

LIFE SKILL

Comparing Expenses

You can use charts and graphs to show information quickly and clearly. To compare parts of a total, or whole, use a circle graph.

The circle graph below shows the expenses for Small Book Company. The whole circle shows the total expenses of the company. The categories shown on the graph represent parts of the whole. By using the graph, you can quickly see the relationship of each category to the total expenses of the company.

To draw a circle graph, follow these steps.

- Write a ratio that compares the parts of the whole to the whole. Write a ratio for each category to be shown on the graph.

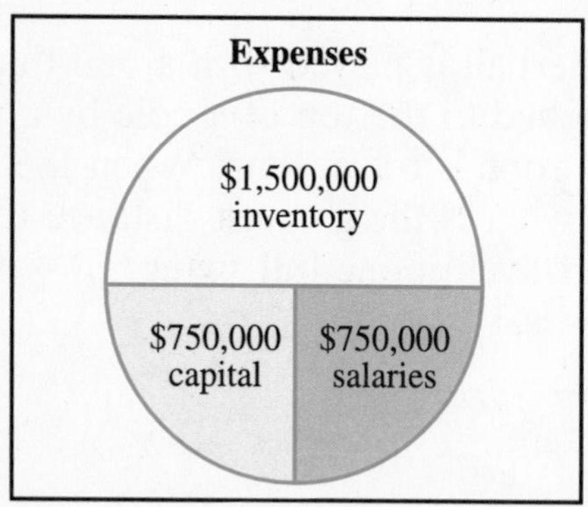

For example, the company's total expenses were $3,000,000. The expense for Inventory was $1,500,000. Set up a proportion to find what percent Inventory was of the total expenses.

$$\frac{1{,}500{,}000}{3{,}000{,}000} = \frac{P}{100}$$

After multiplying diagonals and dividing by the number left, the percent is 50.

- Find the number of degrees of the angle for each part to be shown on the graph.

A complete circle has 360°. To find the number of the degrees of the angle for Inventory, multiply 50% by 360°.

$$360° (0.50) = 180°$$

So, the measurement of the angle for Inventory is 180°.

Complete the chart. Then draw a circle graph that compares the income of the Small Book Company. Total income is $5,000,000.

1.

Categories	Percent	Angle Measure
Sales	40%	________
Royalties	35%	________
Payment for Loans	25%	________

2. How much was earned from Sales?

____

3. How much was earned from Royalties?

____

4. How much was earned from Payment for Loans?

____

LESSON 47

Circles—Areas and Composites

The area of a circle is the surface inside the circle. The formula for finding the area of a circle is $A = \pi r^2$, where r is the radius of the circle.

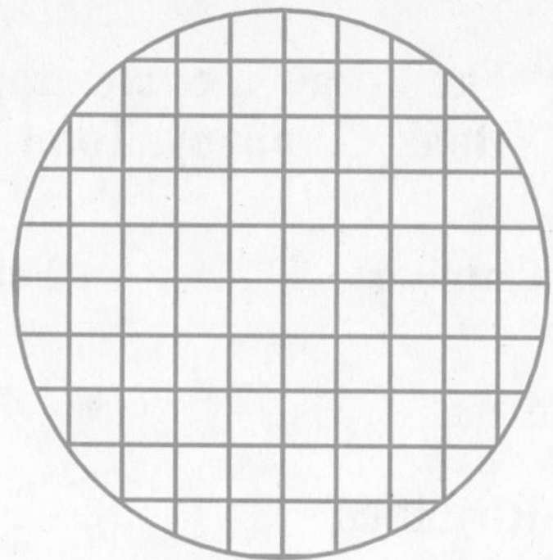

Examples

A. Find the area of a circle with a radius of 8 feet. Use $\pi = 3.14$.

$A = \pi r^2$
$A = 3.14(8)^2$
$A = 3.14(64)$
$A = 200.96$

The area of the circle is about 200 square feet.

Composite figures are complex shapes formed by combining simpler shapes. To find the area of a complex figure, divide the figure into its simplest parts. Then add the areas of the parts to get the total measurement.

B. Find the area of this figure.

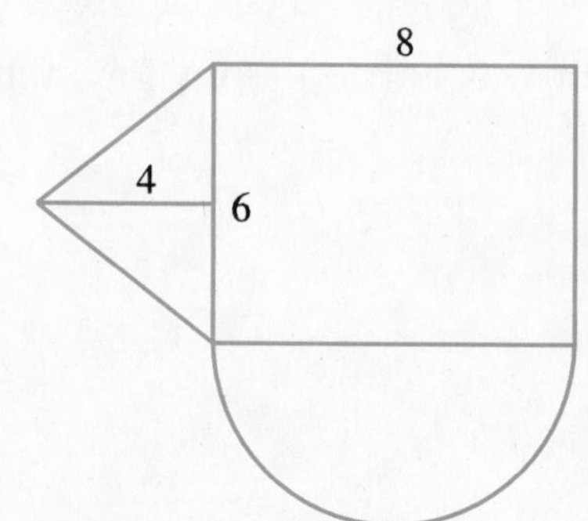

To find the area of the triangle part of the figure, use the formula $A = \frac{1}{2}bh$.

$A = \frac{1}{2}(4)(6)$
$A = 12$

To find the area of the rectangle part, use the formula $A = lw$.

$A = 6(8)$
$A = 48$

 For more information, see Book 5, pages 128–132.

To find the area of the half-circle part, use the formula $A = \frac{1}{2}\pi r^2$.

$A = \frac{1}{2}(3.14)(16)$
$A = 8(3.14)$
$A = 25.12$

The total area is the sum of the parts.

$12 + 48 + 25.12 = 85.12$

The total area of the figure is 85.12.

Practice

Find the area of each figure.

1.

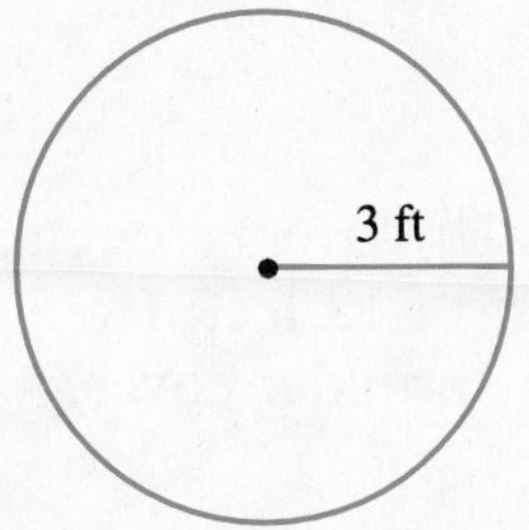

______

2.

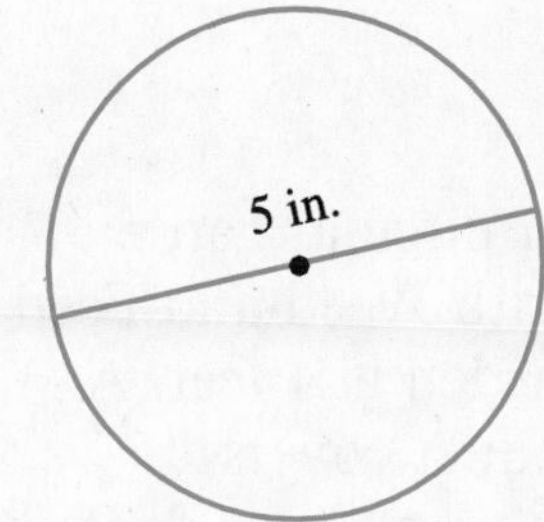

______

3.

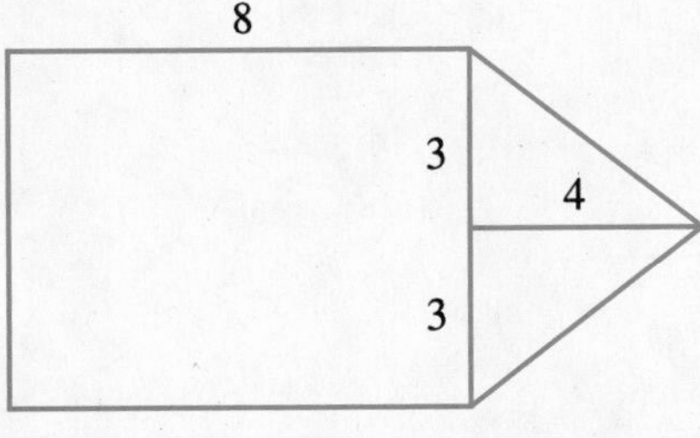

______

4.

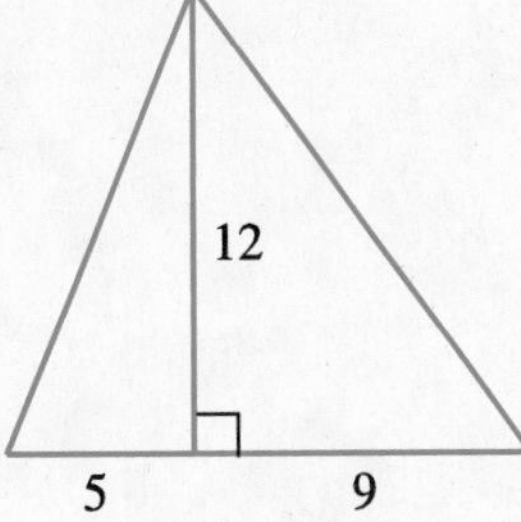

______

5.

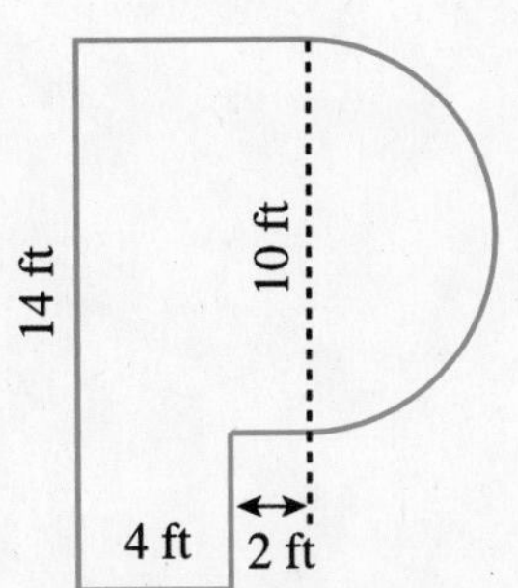

______

Problem Solving

Solve the following problems.

6. The front door of Pierre's house has a window that is a half circle. The window has a diameter of two feet. What is the area of the window?

____

7. Azeez is making a circular tablecloth with a diameter of 3 yards. How much material does he need?

____

8. Thelma fertilizes her circular garden three times a year with Weed-and-Feed. The radius of the garden is 4 feet. A small bag of the product covers 25 square feet. How many bags does she need to cover the garden?

____

9. The Mist-Me sprinkler covers a circular area with a diameter of 12 feet. How much area does the sprinkler cover?

____

Problem Solving—Making a Simpler Problem

The steps you have learned to solve word problems can be used with word problems that deal with geometric figures. Use the following steps:

Step 1 Read the problem and underline the key words. These words will usually relate to some mathematical reasoning.

Step 2 Make a plan to solve the problem. Ask yourself, Should I add, subtract, multiply, divide, round, or compare? You may have to do more than one of these operations for the same problem.

Step 3 Find the solution. Use your math knowledge to find your answer.

Step 4 Check your answer. Ask yourself, Is the answer reasonable? Did you find what you were asked for?

When solving word problems, look for a simpler problem. Breaking a bigger problem into several smaller problems makes your work easier.

Example

The new signs for Main Town Shopping Square are shaped like a rectangle with a half circle on each end. The rectangular section is 1 foot high and 5 feet long. The half circles have 1-foot diameters. How much wood is needed to create each new sign?

Step 1 Read the problem.
The key words are **rectangle** and **circles**.

Step 2 Make a plan to solve the problem.
Find the area of each section of the sign and add the areas together to get the total.

Step 3 Find the solution.
Draw the figure and label the parts.

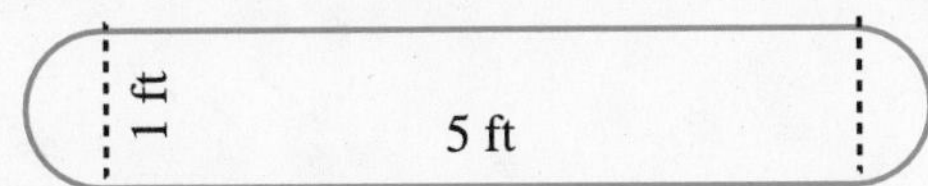

For more information, see Book 5, pages 133–135; 152–154.

The area of the rectangular shape is:

$$A = lw$$
$$A = 1(5)$$
$$A = 5$$

The areas of the half circles are:

$$A = \frac{1}{2}\pi r^2 + \frac{1}{2}\pi r^2$$
$$A = \frac{3.14(0.5)^2}{2} + \frac{3.14(0.5)^2}{2}$$
$$A = 3.14(0.5)^2$$
$$A = 3.14(0.25)$$
$$A = 0.785$$

MATH HINT

The area of 2 half circles is equal to the area of a whole circle.

The total amount of wood needed is 5 + 0.785, or 5.785 square feet.

Step 4 Check your answer.

If you consider the whole sign to be a rectangle, then the area would be $A = 1(6)$, or 6 square feet. The answer is less than 6 square feet, so it is reasonable.

Practice

Draw a diagram for each problem and label it. Then answer each question.

1. A goat is chained by a 10-foot chain to a shed that is 4 feet by 22 feet. The goat is chained to the center of the 22-foot side. Find the amount of grazing area for the goat.

____

2. A circle was cut from a square piece of metal that measured 24 inches square. The circle has a diameter of 24 inches. How much metal was wasted?

____

UNIT

7

Posttest

Identify the following parts of the circle below.

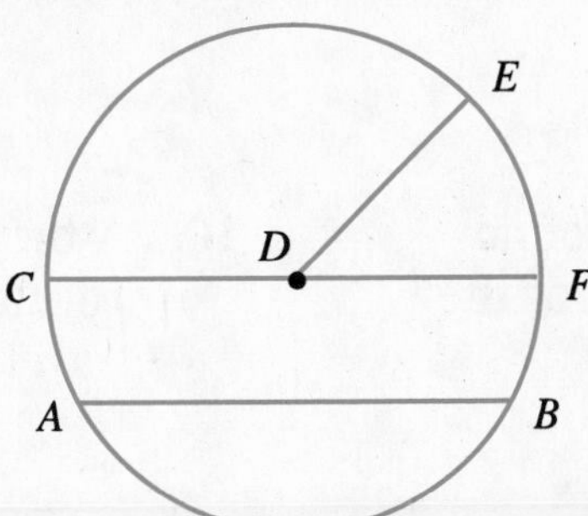

1. radius

______

2. diameter

______

3. center

______

4. chord

______

Find the circumference and area for each circle. Use $\pi = 3.14$.

5.

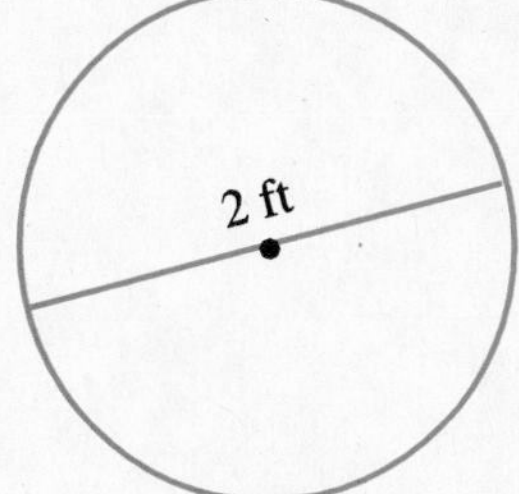

______

6.

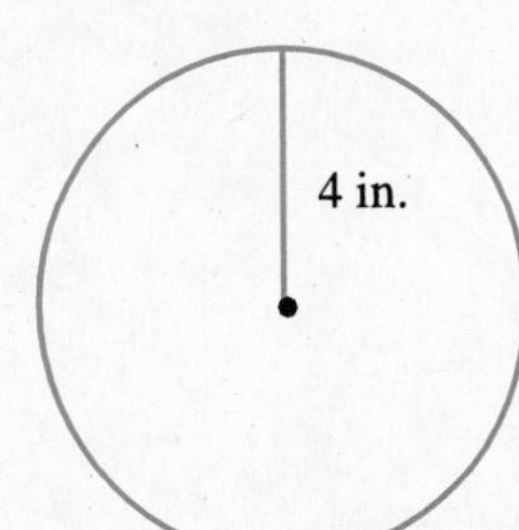

______

7.

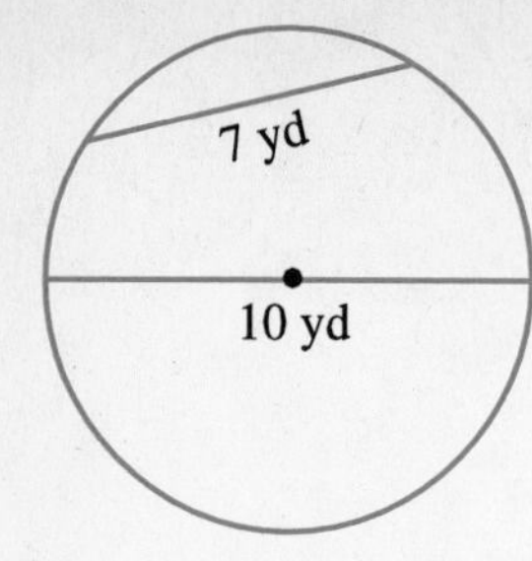

8.

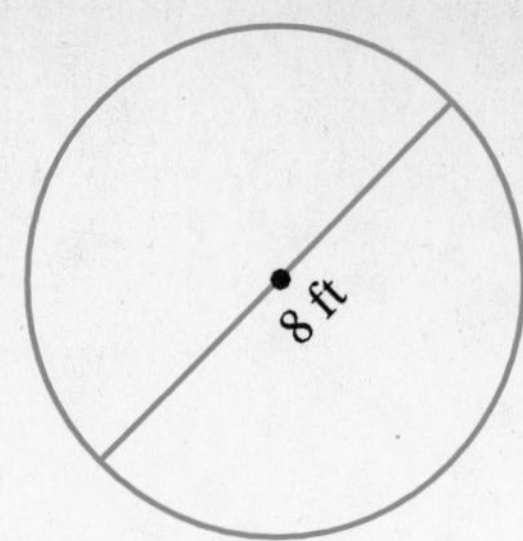

Problem Solving

Solve the following problems.

9. A pizza store sells a 10-inch round pizza for $6. What is the cost of the pizza per square inch?

10. A bicycle wheel has a diameter of 16 inches. How far does the bicycle travel in one revolution of the wheel?

UNIT

8

Surface Areas and Volumes

Pretest

Find the volume and surface area for each of the following figures.

1.

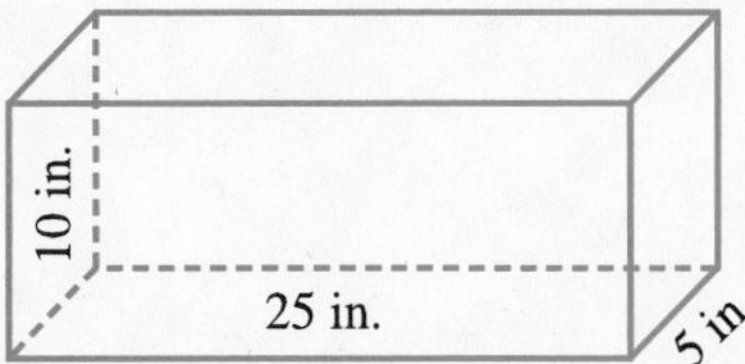

______

______

2.

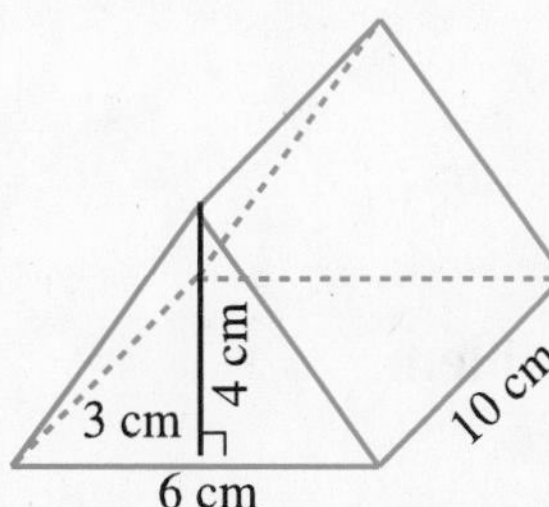

______

______

3.

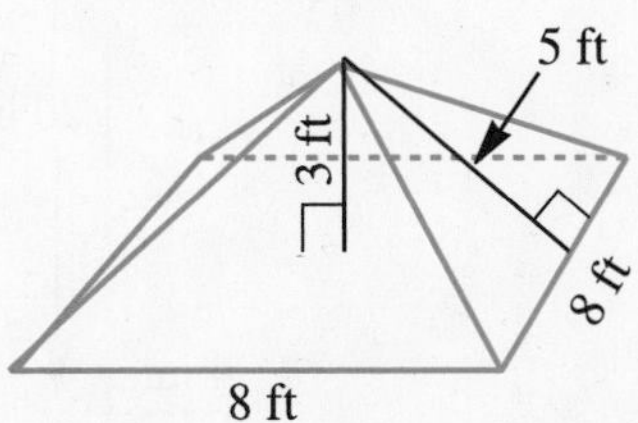

______

______

4.

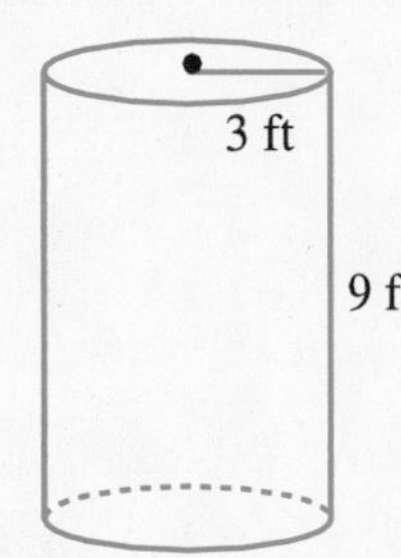

______

______

5.

____

____

6.

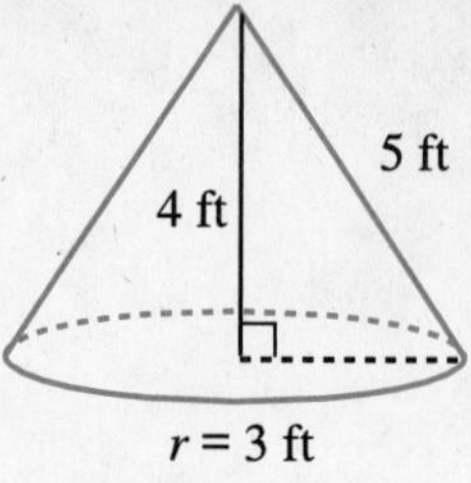

____

____

7.

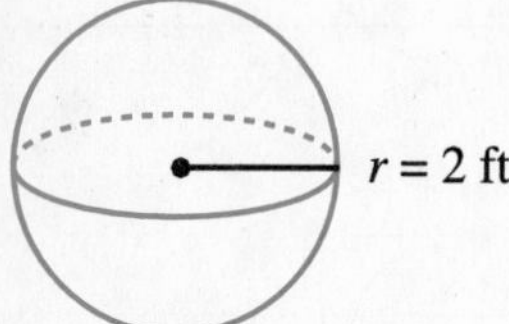

____

____

8.

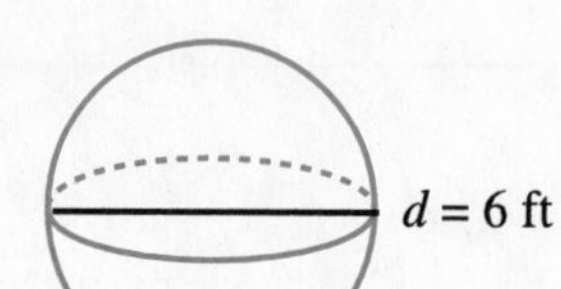

____

____

Problem Solving

Solve the following problem.

9. Squares are cut from each corner of a rectangular piece of cardboard. The dimensions of the cardboard are marked on the figure. If the sides of the cardboard are turned up to make a box, find the volume and surface area of the box created.

____

____

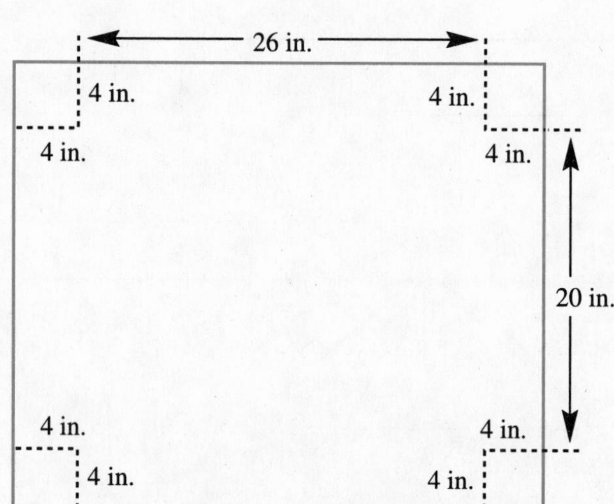

Surface Areas of Prisms and Pyramids

A **prism** is a solid figure with two equal polygons for the bases.

Solid figures have three dimensions: length, width, and height. Solid figures are described by the sides, or faces, of the figures. A **rectangular prism** is a solid figure with all rectangular sides.

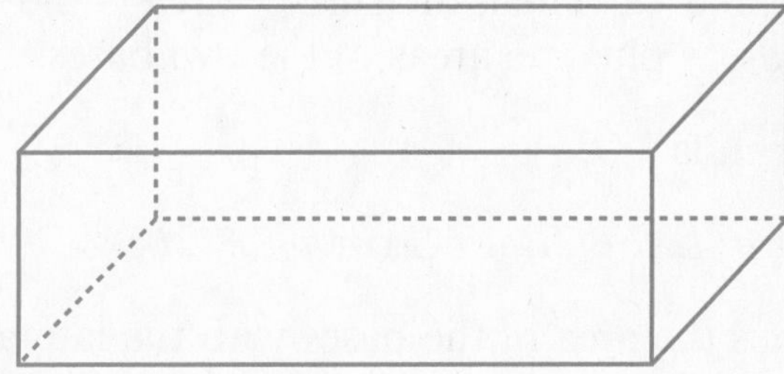

Rectangular prism

The **surface area** of a rectangular prism is the sum of the areas of the faces of the figure.

To find the surface area of a rectangular solid, use the formula

$$S = 2lw + 2wh + 2lh$$

where S stands for surface area, l stands for length, w for width, and h for height.

Example

A. Find the surface area of a rectangular prism that has a length of 8 inches, a width of 12 inches, and a height of 6 inches.

$S = 2lw + 2wh + 2lh$
$S = 2(8)(12) + 2(12)(6) + 2(8)(6)$
$S = 192 + 144 + 96$
$S = 432$

The total surface area is 432 in^2.

The bases of a prism are opposite each other. The corresponding vertices of the bases are connected by line segments. A prism is named for the figure that forms the base. The figure at right is a **triangular prism**. The bases are triangles.

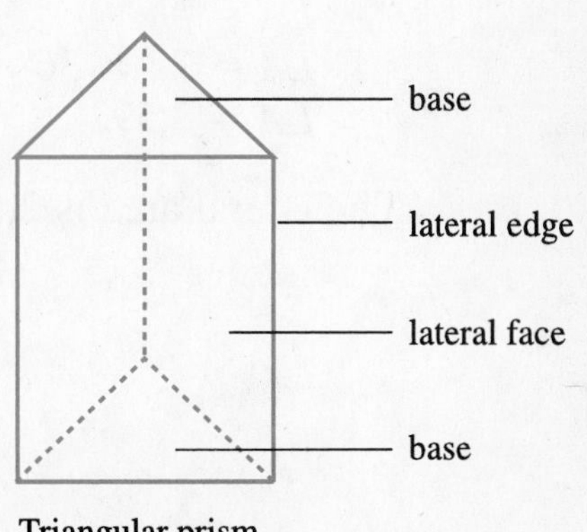

Triangular prism

For more information, see Book 5, pages 197–200.

The **lateral area of a prism** is the area of just the sides of a prism. To find the lateral area of a prism, use the formula

$$LA = ph$$

where LA is the lateral area, p is the perimeter of the base, and h is the height of the prism.

Another way of finding the **surface area of a prism** is to use the lateral area. Then, the surface area is the sum of the lateral areas plus the areas of the two bases.

The formula

$$SA = LA + 2B, \text{ or } SA = ph + 2B$$

combines the area of the bases with the lateral area formula.

MATH HINT

Remember:
p = perimeter of the base
h = height of the prism
B = area of the base
LA = lateral area
SA = surface area

Example

B. Find the surface area of a hexagonal prism that has sides measuring 12 inches and a height of 36 inches; the area of the base is 375.84 in².

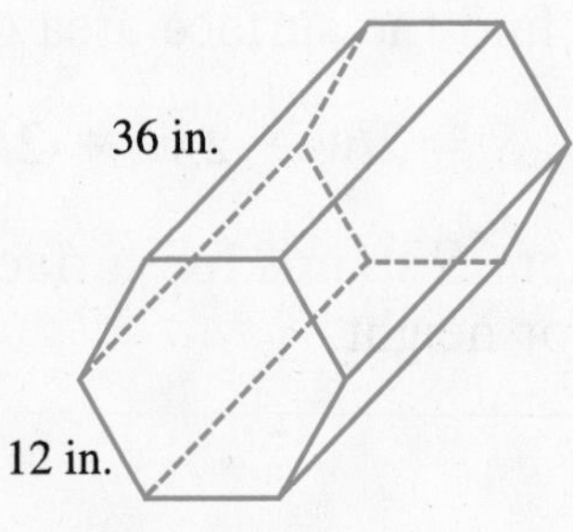

Hexagonal prism

Step 1 Find the lateral area of the prism.

$LA = ph$

To find the perimeter of the base, multiply.

$$6 \times 12 = 72$$

The perimeter of the base is 72 inches.

$$LA = ph$$
$$LA = 72 \times 36$$
$$LA = 2{,}592$$

The lateral area is 2,592 in².

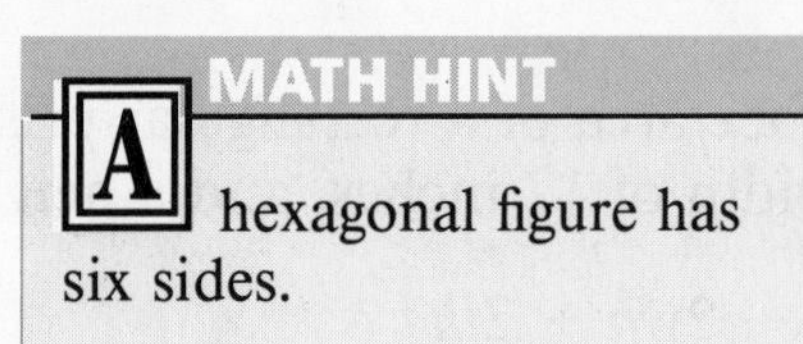
MATH HINT

A hexagonal figure has six sides.

Step 2 Find the surface area of the prism.
The area of the base is 375.84 in^2.

$SA = LA + 2B$
$SA = 2{,}592 + 2(375.84)$
$SA = 2{,}592 + 751.68$
$SA = 3{,}343.68$

The surface area is 3,343.68 in^2.

A **pyramid** is a solid figure with a polygon for a base and triangles for faces. A pyramid is named for the shape of its base. Common pyramids have bases that are triangles, rectangles, pentagons, and hexagons.

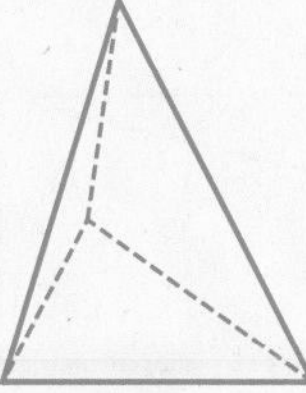
Triangular pyramid

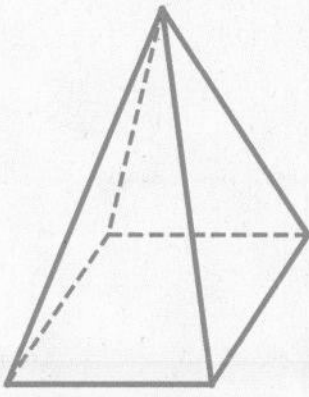
Rectangular pyramid

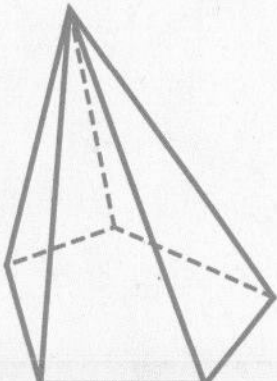
Pentagonal pyramid

The **surface area of a pyramid** is found in the same way as the surface area of a prism. It is the sum of the area of the base and the areas of the lateral faces. Since the lateral faces are triangles, the lateral area, *LA,* is found by the formula:

$LA = \frac{1}{2}pl$

where p is the perimeter of the base and l is the height of the lateral faces.

$SA = LA + B$

Example

C. The figure shown is a pyramid. Find its surface area.

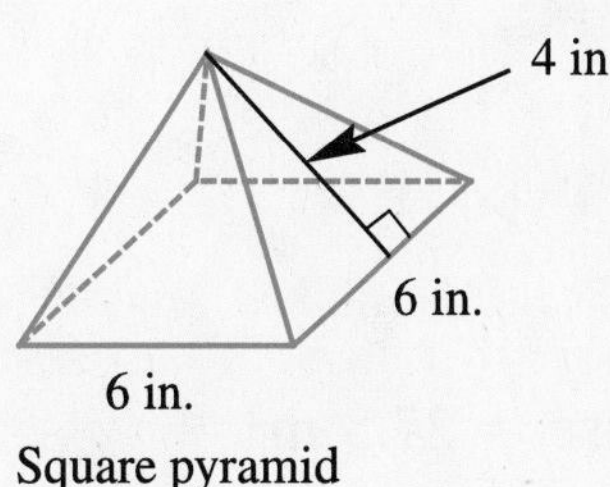

Square pyramid

Step 1 Find the lateral area of the pyramid.

$LA = \left(\frac{1}{2}\right)pl$
$LA = \left(\frac{1}{2}\right)(24)(4)$
$LA = 48$

The lateral area is 48 in^2.

Step 2 Find the surface area of the pyramid. The area of the base is 6(6), or 36 in².

$SA = LA + B$
$SA = 48 + 36$
$SA = 84$

The surface area is 84 in².

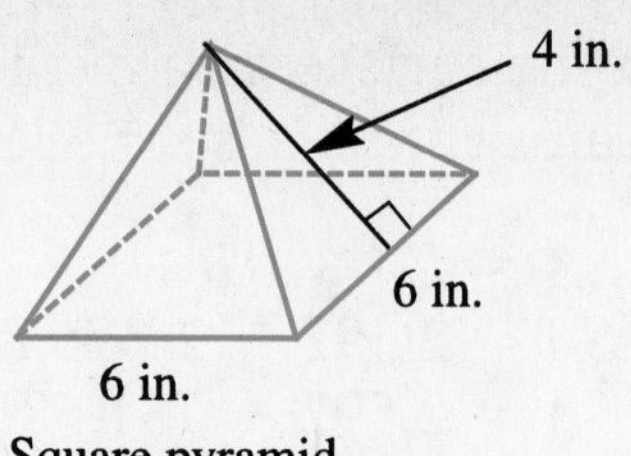

Square pyramid

Practice

Find the total surface area for each of the following figures.

1.

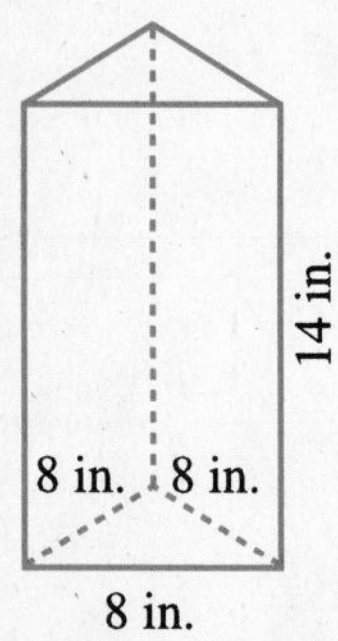

Area of Base = 27.72 in²

____

2.

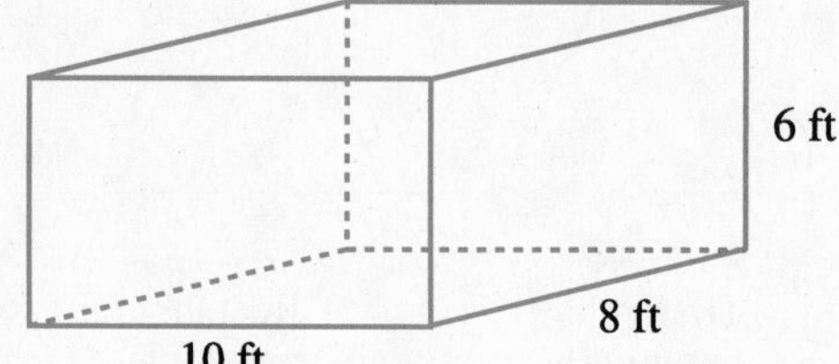

____

3.

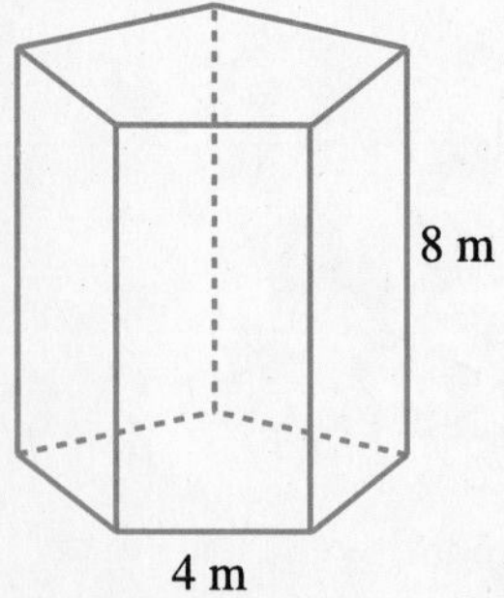

Area of Base = 27.5 m²

____

4.

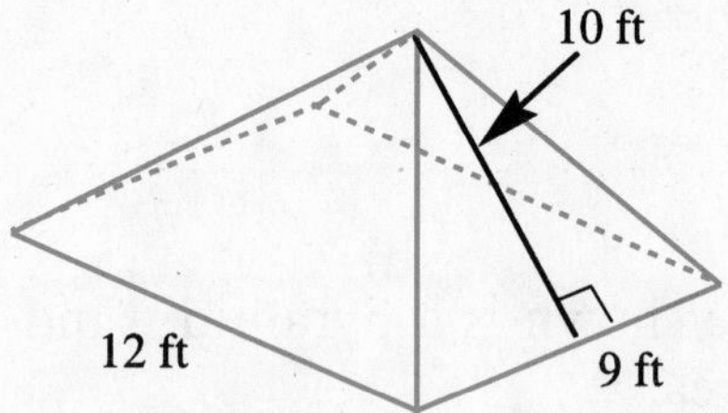

____

Volumes of Prisms and Pyramids

Volume is a measure of capacity. It tells you how much space is inside a solid and is measured in cubic units. The volume of a prism is equal to the area of its base times its height. Use this formula

$V = Bh$

where B is the area of the base and h is the height.
For a rectangular prism, the formula is $V = Bh$, or $V = lwh$, where l is the length, w is the width, and h is the height.

Examples

A. Find the volume for the rectangular solid with a length of 12 inches, a width of 8 inches, and a height of 6 inches.

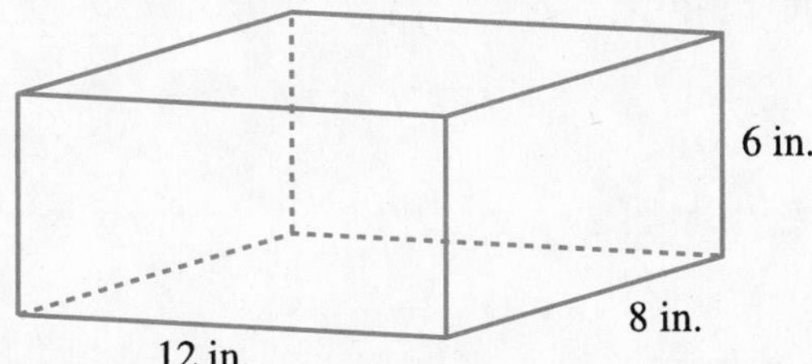

The base is a rectangle. To find the area of the base, use the formula:

$A = lw$
$A = (8)(12)$
$A = 96$

The area of the base is 96 in^2.

To find the volume of the rectangular solid, use the formula

$V = Bh$
$V = 96(6)$
$V = 576$

The volume of the figure is 576 in^3.

For more information, see Book 5, pages 205–206.

B. Find the volume of the square prism.

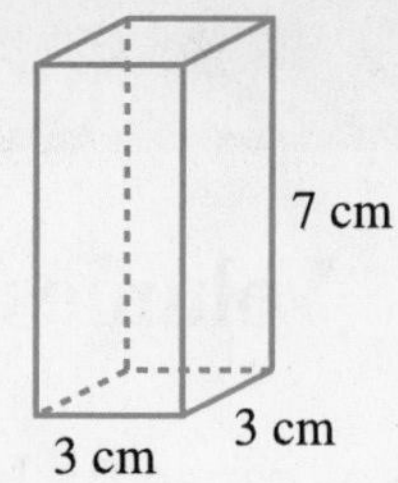

The base is a square. To find the area of the base, use the formula:

$A = s^2$
$A = 3^2$
$A = 9$

The area of the base is 9 cm^2.

To find the volume of the square prism, use the formula:

$V = Bh$
$V = 9(7)$
$V = 63$

The volume of the figure is 63 cm^3.

The volume of a pyramid is one-third the area of the base times the height. Use the formula

$V = \frac{1}{3}Bh$

where B is the area of the base and h is the height.

Example

C. Find the volume of the regular square pyramid.

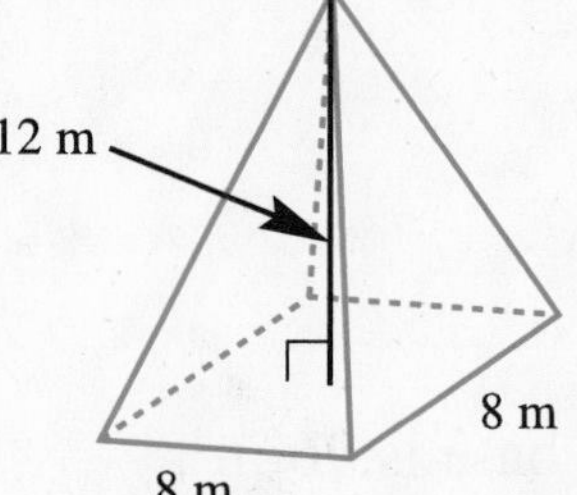

The base is a square, so use the formula $A = s^2$.

$A = (8)(8)$, or 64

The area of the base is 64 cm^2.

To find the volume of the pyramid, use the formula:

$V = \frac{1}{3}Bh$
$V = \frac{1}{3}(64)(12)$
$V = 256$

The volume of the pyramid is 256 m^3.

Practice

Find the volume of each figure below.

1.

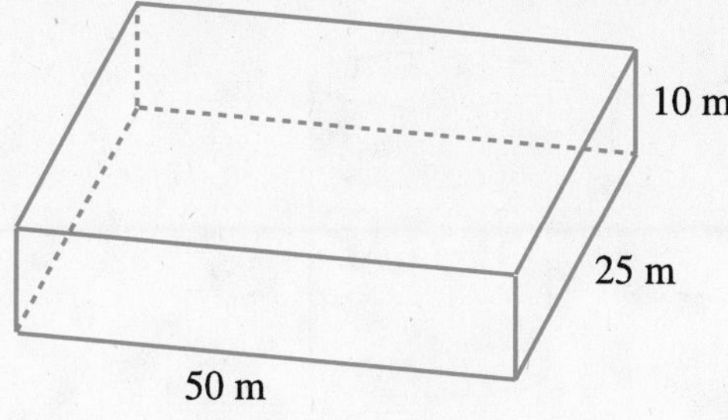

____

2.

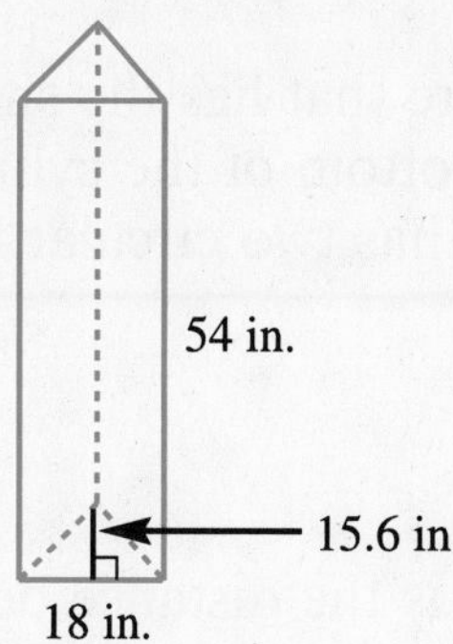

____

3.

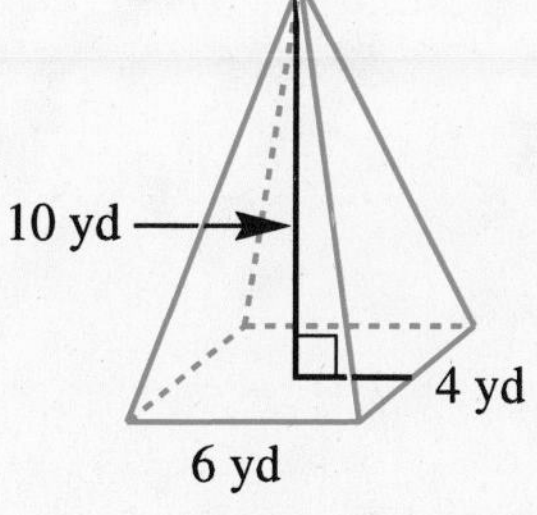

____

4.

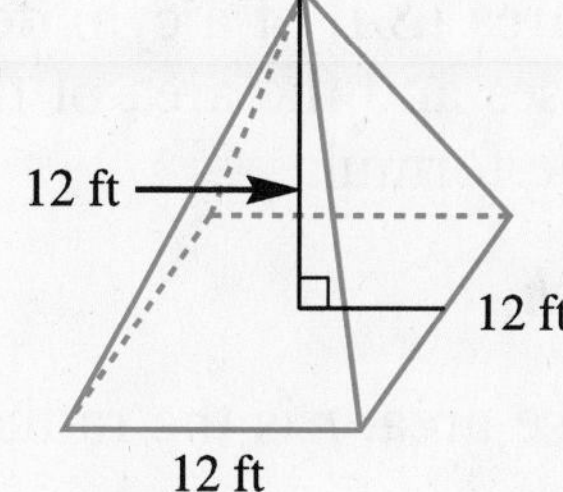

____

Problem Solving

Solve the following problem.

5. A company wants to package small cereal boxes. If each cereal box measures 3 inches long, 2 inches deep, and 4 inches high, what would be the total surface area of 4 boxes placed next to each other?

____

Cylinders and Cones

A **cylinder** is a solid figure that has the shape similar to a soup can. The top and bottom of the cylinder are called bases. A cylinder always has two circular bases.

The height of a cylinder is the distance between the bases. The area of the label part of the can is the lateral surface area (*LA*). The formula to find this area is

$$LA = 2\pi rh$$

where *LA* is the lateral surface area, *r* is the radius, and *h* is the height. Total surface area (*SA*) of a cylinder is the sum of the areas of the two circular bases and the area of the rectangle that forms its lateral surface. Use the formula

$$SA = 2\pi r^2 + 2\pi rh$$

where *SA* is the surface area, *r* is the radius, and *h* is the height.

To find the volume of a cylinder, use the formula

$$V = \pi r^2 h$$

where *V* is the volume, *r* is the radius, and *h* is the height.

Example

A. Find the surface area and the volume of a cylinder with the radius of 5 feet and a height of 4.5 feet.

Surface Area

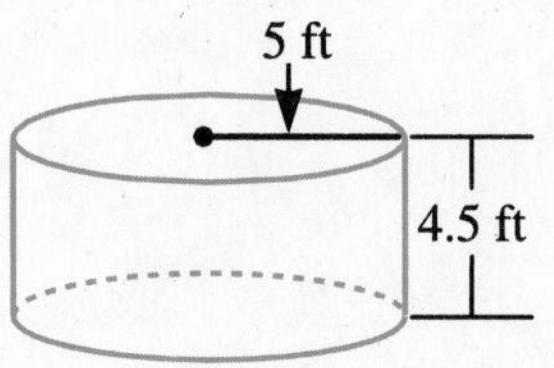

$$SA = 2\pi r^2 + 2\pi rh$$
$$SA = 2(3.14)(5)^2 + 2(3.14)(5)(4.5)$$
$$SA = 2(3.14)(25) + 2(3.14)(5)(4.5)$$
$$SA = 157 + 141.3$$
$$SA = 298.3 \text{ ft}^2$$

 For more information, see Book 5, pages 178–179.

Volume

$$V = \pi r^2 h$$
$$V = 3.14(5)^2(4.5)$$
$$V = 3.14(25)(4.5)$$
$$V = 353.25 \text{ ft}^3$$

A **cone** is a solid figure with a circular base and the set of all possible line segments joining a point not on the circle to the base. A common example of a cone is an ice cream cone. Many party hats are shaped like cones.

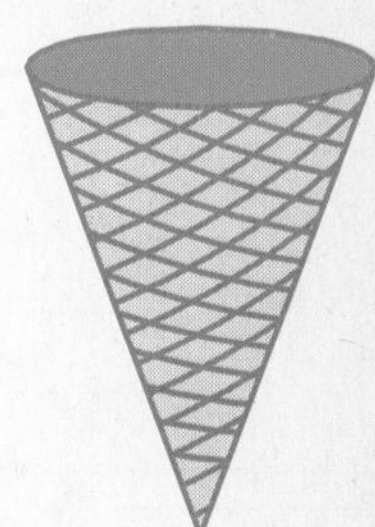

The total surface area is the lateral surface area plus the surface area of the base. To find the lateral surface of a cone use the formula

$$LA = \pi rs$$

where LA is the lateral surface area, s is the height, and r is the radius.

To find the total surface area of a cone, use the formula

$$SA = \pi rs + \pi r^2$$

where SA is the surface area, s is the height, and r is the radius.

The **slant height** is the line segment joining the vertex to the base. To find the slant height, use the Pythagorean Theorem.

$s = \sqrt{r^2 + h^2}$ where r is the radius and h is the height.

To find the volume of a cone, use the formula

$$V = \frac{1}{3}\pi r^2 h$$

where V is volume, h is the height, and r is the radius.

Example

B. Find the surface area and the volume for a cone with a diameter of 8 inches and a height of 9 inches.

To find the surface area, follow these steps:

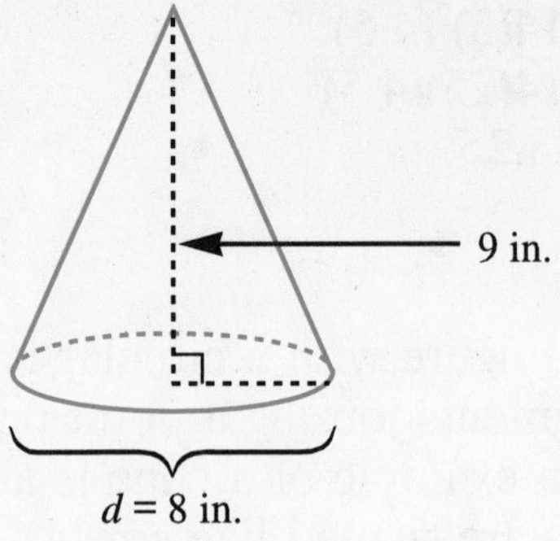

Step 1 Find the slant height.

$s = \sqrt{r^2 + h^2}$
$s = \sqrt{4^2 + 9^2}$
$s = 9.8489$ inches

Step 2 Find the lateral area.

$LA = \pi rs$
$LA = 3.14(4)\ (9.8489)$
$LA = 123.70218$

The lateral area is about 123.70 in^2.

Step 3 Find the surface area.

$SA = \pi r^2 + \pi rs$
$SA = 3.14(16) + 123.7$
$SA = 173.94$

The surface area is 173.94 in^2.

To find the volume, do the following:

$V = \frac{1}{3}\pi r^2 h$
$V = \frac{1}{3}(3.14)(4)^2(9)$
$V = \frac{1}{3}(3.14)(16)(9)$
$V = 150.72$

The volume is 150.72 in^3.

Practice

Find the volume, lateral surface area, and the total surface area for each of the following cylinders.

1. radius = 10 ft
height = 8 ft

______
______
______

2. radius = 3.1 cm
height = 19.2 cm

______
______
______

Find the volume, lateral surface area, and the total surface area for each of the following cones.

3. radius = 6 cm
height = 12 cm

______
______
______

4. radius = 5 m
height = 18 m

______
______
______

Problem Solving

Solve the following problems.

5. Joe stores grain in a 60-foot cylinder with a 24-foot diameter. How many bushels of grain can he store? (Hint: 1 bushel = 1.25 ft^3)

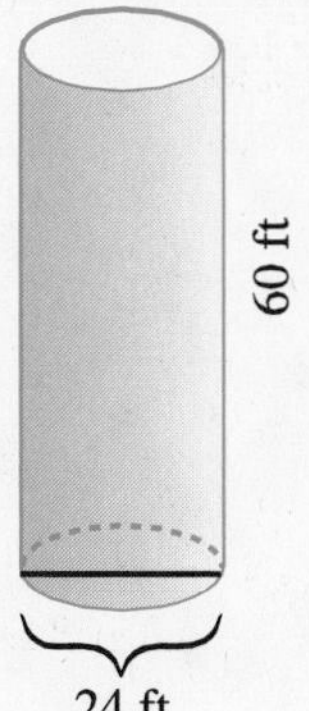

______

6. An ice cream shop sells a pint of ice cream for $2.50. The volume of the pint of ice cream is 28.875 in^3. Find the cost per cubic inch for a pint of ice cream.

______

LIFE SKILL

Water Rates

In some cities, you pay for the water you use each month. Some areas bill customers every three months. Sometimes the water bill includes other items such as trash pick-up and waste water disposal.

In any case, water is charged by the amount you use. The charges for water are usually determined at a rate per 100 cubic feet. Suppose the rate for your area is $1.50 per 100 cubic feet of water.

What would it cost you to bathe every day for one month? Assume that you have a rectangular bathtub.

An average rectangular bathtub measures 4 feet by 2 feet by 1 foot.

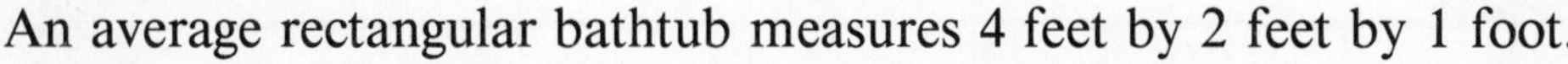

To find the number of cubic feet of water it takes to fill the tub, use the formula

$$V = lwh$$
$$V = 4(2)(1)$$
$$V = 8$$

It takes 8 cubic feet of water to fill the bathtub.

Since the rate for 100 cubic feet of water is $1.50, the rate for one cubic foot of water would be $0.015.

The cost to fill the bathtub would be 8(0.015), or $0.12.

So, if you take a bath every day for a month, it would cost you 30(0.12), or $3.60.

Other household conveniences, such as automatic clothes washers and dishwashers, use large amounts of water. Each time you use one of these items, you are using water and spending money. You can save money and conserve water by reducing the amount of water used by these conveniences.

In the following problem, determine the cost in one month using the above rate.

A clothes washer uses 5 cubic feet of water during the wash cycle and another 5 cubic feet of water during the rinse cycle. What is the cost for one month if you do four loads of laundry a week? Remember that the rate per cubic foot is $0.015. ____________

Spheres

A **sphere** is a solid with all points the same distance from the center.

A common example is a baseball. Earth also looks like a sphere. A sphere has a center point, a diameter, and a radius.

To find the surface area of a sphere, use the formula

$$SA = 4\pi r^2$$

where SA is surface area and r is the radius of the sphere.

To find the volume of a sphere, use the formula

$$V = \frac{4}{3}\pi r^3$$

where V is the volume and r is the radius of the sphere.

 For more information, see Book 5, pages 216–223.

Example

Find the surface area and the volume of a sphere with a radius of 12 meters.

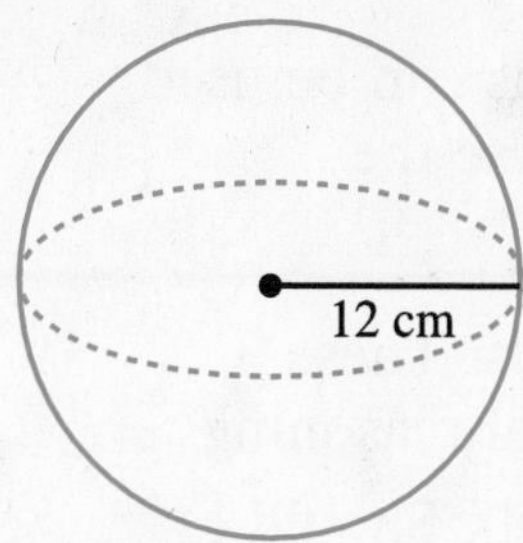

Surface Area

$SA = 4\pi r^2$
$SA = 4(3.14)(12)^2$
$SA = 4(3.14)(144)$
$SA = 1{,}808.64$

The surface area is 1,808.64 m^2.

Volume

$V = \left(\frac{4}{3}\right)\pi r^3$
$V = \left(\frac{4}{3}\right)(3.14)(12)^3$
$V = \left(\frac{4}{3}\right)(3.14)(1{,}728)$
$V = 7{,}234.56$

The volume is 7,234.56 m^3

Practice

Find volume and surface area for each of the following spheres. (Hint: Recall that $r = \frac{1}{2}d$.)

1. Golf ball; $d = 1$ inch

__________ __________

2. Tennis ball; $d = 2.5$ inches

__________ __________

3. Basketball; $d = 9$ inches

__________ __________

4. Earth; $d = 7{,}926$ miles

__________ __________

5. Moon; $d = 2{,}160$ miles

__________ __________

6. Sun; $d = 864{,}000$ miles

__________ __________

Problem Solving—Drawing a Picture

The steps you have learned to help solve word problems can be used with word problems that deal with geometric figures. Use the following steps:

Step 1 Read the problem and underline the key words. These words will usually relate to some mathematical reasoning.

Step 2 Make a plan to solve the problem. Ask yourself, Should I add, subtract, multiply, divide, round, or compare? You may have to do more than one of these operations for the same problem.

Step 3 Find the solution. Use your math knowledge to find the answer.

Step 4 Check your answer. Ask yourself, Is the answer reasonable? Did you find what you were asked for?

It is helpful to draw the figures that are described in a problem. Making the drawing allows you to see what you know.

Example

A cylinder is used to store oil. The cylinder is 24 feet in diameter and 60 feet tall. If one cubic foot of oil equals 7.48 gallons, how many gallons of oil can the cylinder hold?

Step 1 Read the problem and identify the key words. The key words are **how many gallons**, **diameter**, and **tall**.

Step 2 Make a plan to solve the problem. Draw the cylinder. Then find the volume of the cylinder, and multiply the results by 7.48 to find the number of gallons stored.

Step 3 Find the solution.

$V = \pi r^2 h$
$V = 3.14(12)^2(60)$
$V = 27{,}129.6$ cubic feet
$27{,}129.6(7.48) = 202{,}929$ gallons

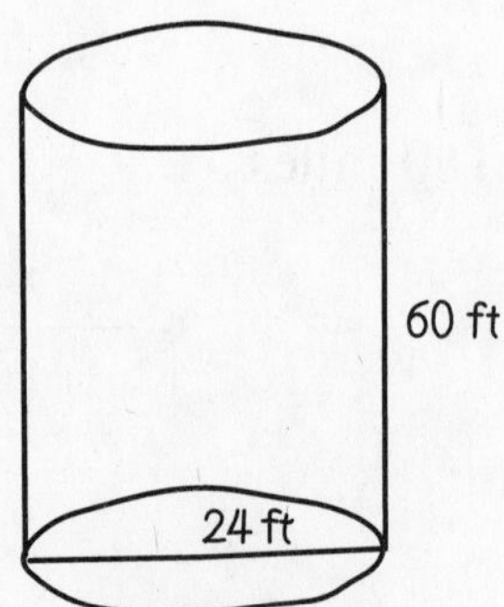

The cylinder can hold 202,929 gallons of oil.

 For more information, see Book 5, pages 207–209.

Step 4 Check your answer.

$$\frac{202{,}929}{7.48} = 27{,}129.6$$

$$\frac{27{,}129.6}{(3.14)(60)} = 144$$

The answer checks working backwards.
The answer is reasonable.

Problem Solving

Solve the following problems.

1. A cylinder measures 15 feet in diameter and is 40 feet high. If 1 bushel of grain can be stored in 1.25 cubic feet, how many bushels of grain can be stored in the cylinder? ______________

2. A silo is shaped like a cylinder with a half sphere on its top. The silo is 20 feet high and has a radius of 7 feet. Find the surface area of the silo.

____

3. Jim and Karen are building a doghouse that measures 4 feet in length by 3 feet wide by 2 feet high. They plan to paint the outside of the doghouse with a paint that covers 5 square feet per pint. If they leave one end of the house open, how many pints of paint do they need? ______________

UNIT

8

Posttest

Find the volume and surface area for each of the following figures.

1.

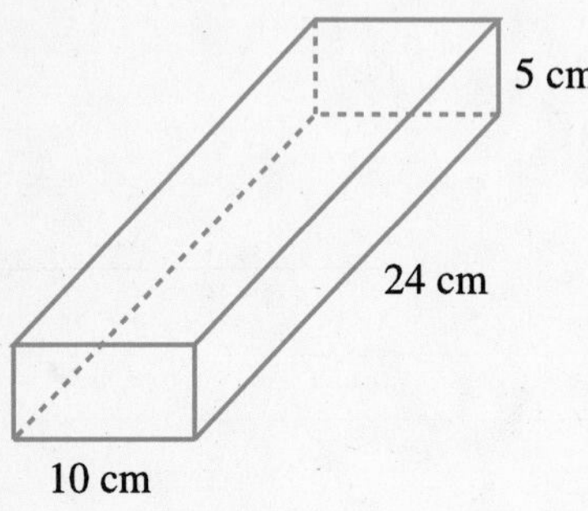

______

______

2.

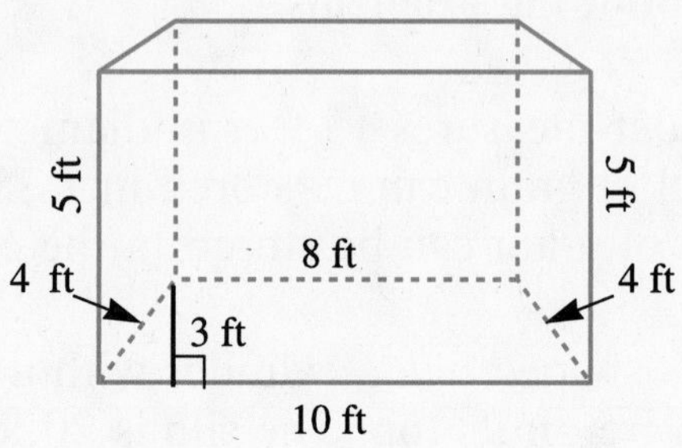

______

______

3.

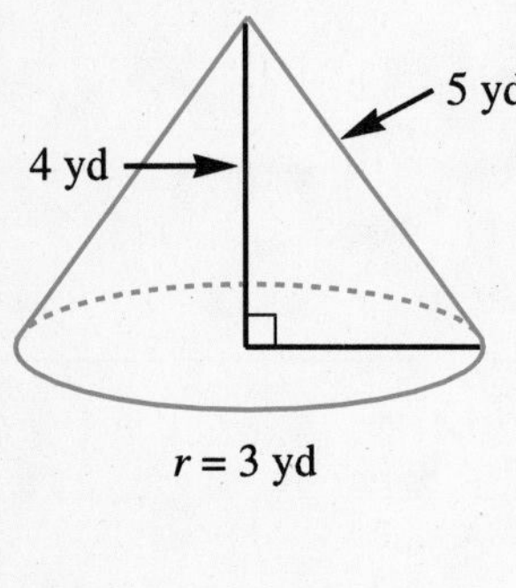

______

______

4.

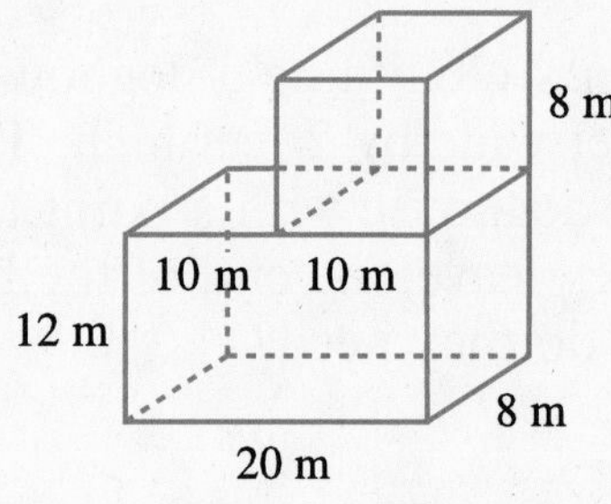

______

______

5.

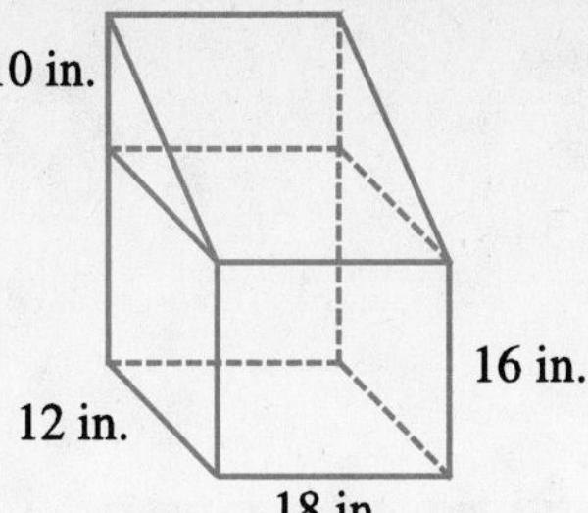

6.

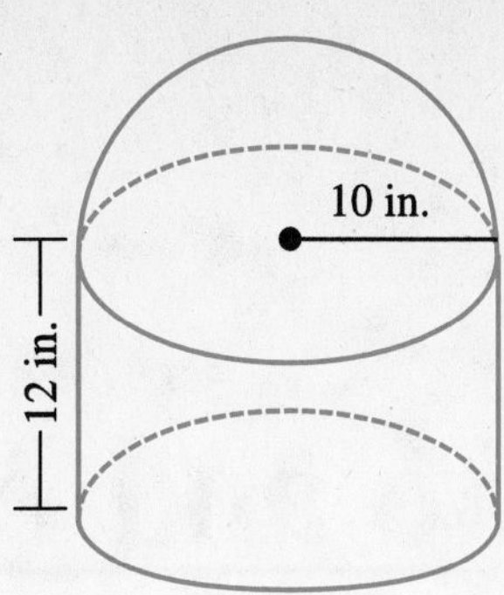

Problem Solving

Solve the following problem.

7. A cone-shaped party hat is filled with popcorn. The party hat has an 8-inch diameter and is 8 inches tall. How much popcorn can it hold?

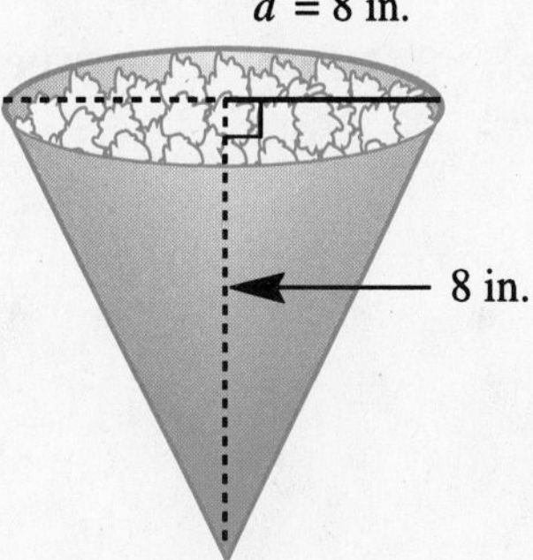

UNIT

9

Positive and Negative Numbers

Pretest

Place the correct symbol, <, >, or =, between the two numbers to make a true statement.

1. 8 ____ -8 **2.** 0 ____ 3 **3.** (-3) ____ -5

4. $-(5)$ ____ 3 **5.** $3(11)$ ____ $-3(-11)$ **6.** -5 ____ 5

Add or subtract.

7. $6 + (-4)$ ____ **8.** $-1 + (-26)$ ____ **9.** $(-21) + 44$ ____

10. $23 + (-18)$ ____ **11.** $3 + (-3)$ ____ **12.** $6 - \left(-\frac{1}{5}\right)$ ____

13. $13 - 15\frac{1}{4}$ ____ **14.** $28 - 57.3$ ____ **15.** $18 - (-18)$ ____

Multiply or divide.

16. $8(-7)$ ____ **17.** $(-4)(12)$ ____ **18.** $(13)(-5)(-1)$ ____

19. $\frac{(2)}{(-3)}$ ____ **20.** $35\left(\frac{-1}{35}\right)$ ____ **21.** $\frac{12}{-4}$ ____

22. $(-6)(-3)(-7)$ __________

23. $(-15)(28)(-2)$ __________

24. $(5)(11)[(-2-(-15)]$ __________

25. $\frac{-80}{0}$ __________

26. $9-3(4^2+2)+5^2-20$ __________

27. $\frac{-4(2+3^2)}{6-2(2^2-2)}$ __________

Problem Solving

28. The water of a river rises and falls during the spring as the ice and snow melt. The levels of a river were checked over a six-week period with the following results. Determine the net change of water level. __________

Week	Water Level Above or Below the Normal
1	+10 feet
2	−5 feet
3	+8 feet
4	+4 feet
5	−10 feet
6	−15 feet

Reading and Writing Positive and Negative Numbers

Positive and negative numbers are called **signed numbers**. You can use a number line to locate signed numbers. **Positive** numbers are shown to the right of zero and **negative** numbers are shown to the left of zero. Zero is neither positive nor negative.

Signed numbers extend indefinitely in both directions on the number line. The numbers get larger as you move to the right and smaller as you move to the left.

When comparing numbers, use the > symbol to mean "greater than" and the < symbol to mean "less than."

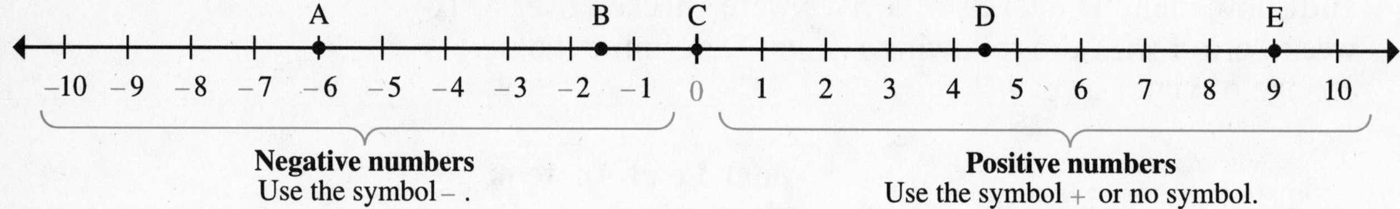

Examples

A. Write the signed number for each letter on the number line above.

Point A is −6 because it is 6 units to the left of zero.
Point B is −1.5 because it is 1.5 units to the left of zero.
Point C is 0 because it is 0 units from zero.
Point D is 4.5 because it is 4.5 units to the right of zero.
Point E is 9 because it is 9 units to the right of zero.

When using a number line to compare two numbers, locate the numbers on the number line. Then look to see which number is to the right or left. Write the relationship using an inequality symbol.

B. Use the number line above to compare Points A and B. Use < to show "less than" or > to show "greater than."

A ______ B   A is to the left of B, so $A < B$.
B ______ A   B is to the right of A, so $B > A$.

MATH HINT

An inequality symbol always "points" to the smaller number.

 For more information, see Book 6, pages 46–47.

C. Is -5 less than or greater than -3?

-5 is the smaller number.
$-5 < -3$, or negative 5 is less than negative 3.

MATH HINT

All negative numbers are less than any positive number. All positive numbers are greater than any negative number.

Each number on the number line has an opposite. Opposite numbers are the same distance from zero in opposite directions on the number line.

MATH HINT

The symbol $-$ can mean subtraction: $8 - 5$. It can also mean a negative number, such as -7.

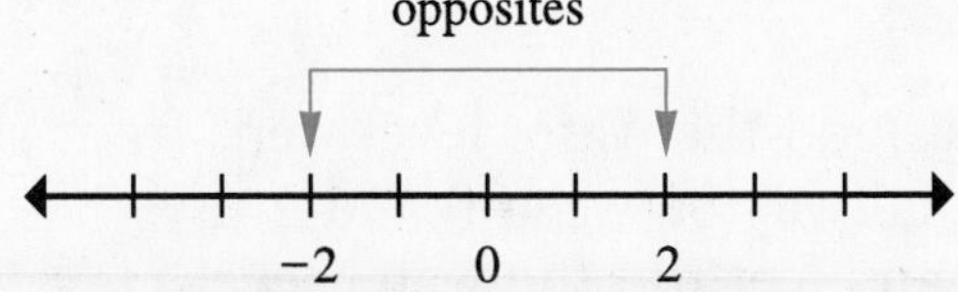

The opposite of 2 is negative 2.

$-(2) = -2$

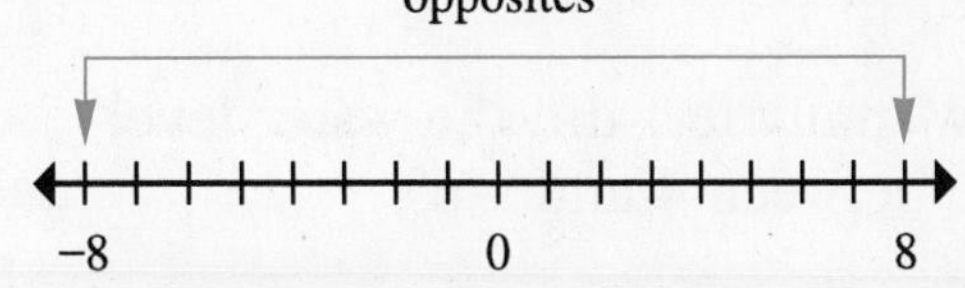

The opposite of negative 8 is 8.

$-(-8) = 8$

Practice

Write whether each statement is true or false.

1. $-6 < 0$ _____
2. $9 > -5$ _____
3. $-3 > -7$ _____
4. $0 > -\frac{1}{2}$ _____
5. $7 > -7$ _____
6. $-16 < -12$ _____

Simplify.

7. $-(10)$ _____
8. $-(-1)$ _____
9. $-(-13)$ _____
10. $-(15 + 3)$ _____
11. $-(3 - 2)$ _____
12. $-(20 + 13)$ _____

List the values from smallest to largest.

13. $-5, -2, 0, -(-1)$

____

14. $-(2), -(-4), 3, -(-5)$

____

Problem Solving

Read the information below. Solve the problems that follow.

Dams and reservoirs have numbers that show present water levels in terms of the number of feet above or below the normal water level. The most visible number at the bottom indicates the current reservoir water level. When the water level is at 0, the reservoir is at its normal water level.

15. After a storm, the water level is at 4. What does this tell us about the water level?

____

16. After a hot dry spell, the water level dropped to −6. What does this tell us about the water level?

____

17. By how many feet did the water level change between 4 and −6?

____

18. After a flood the water level went from −6 to halfway between 0 and 2. By how many feet did the water level change?

____

Use the chart below to answer problems 19–22.

World Location	Distance Above or Below Sea Level
Mt. Everest in Nepal/Tibet Mt. McKinley in Alaska	29,028 feet 30,320 feet
Dead Sea in Israel/Jordan	−1,312 feet
Death Valley in California	−282 feet
Mariana Trench in the Pacific	−36,198 feet

19. Which is lower, the Dead Sea or Death Valley?

____

20. Which is higher, Mt. McKinley or Mt. Everest?

____

21. Which is farther from sea level, Mt. Everest or the Mariana Trench?

____

22. Which location is closest to sea level?

____

Adding and Subtracting Positive and Negative Numbers

Adding positive and negative numbers is the same as moving along a number line.

Find the sum of 5 and 2.

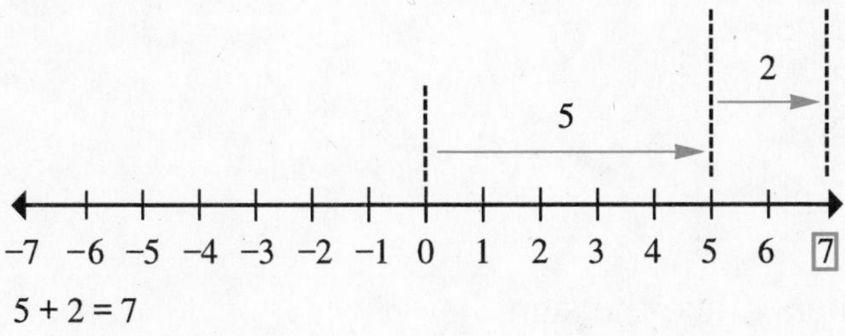

Find the sum of −5 and −2.

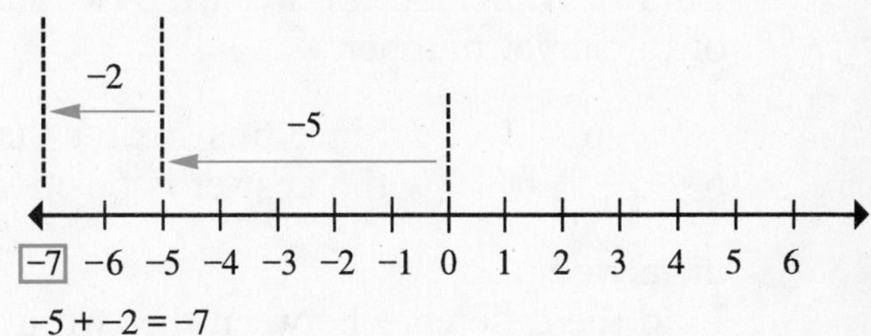

Find the sum of 5 and −2.

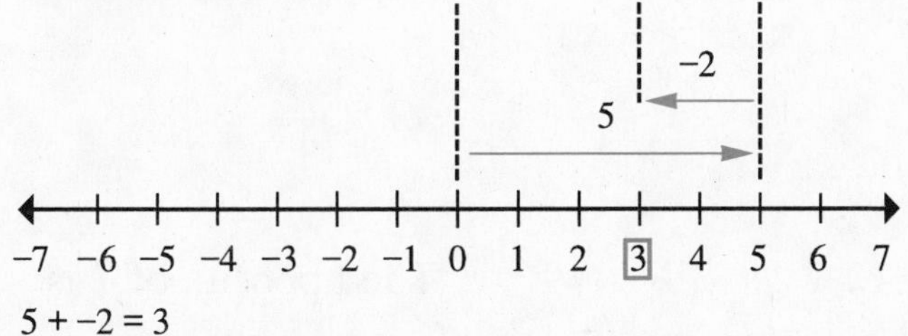

Find the sum of −5 and 2.

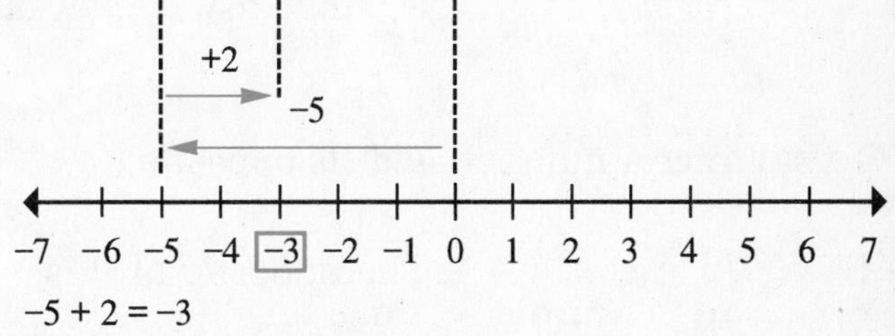

Find the sum of 5 and −5.

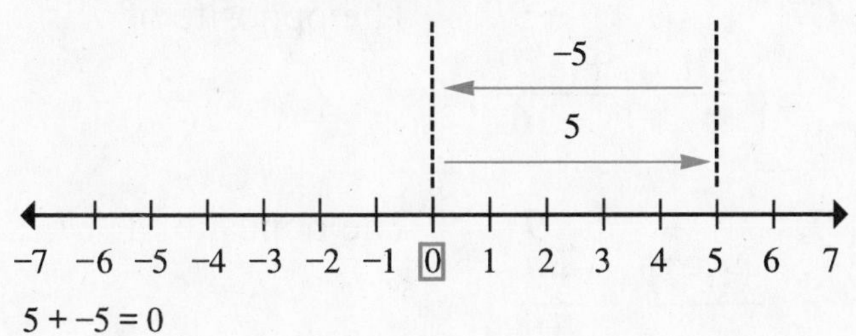

MATH HINT

Moving to the right is positive; moving to the left is negative.

Examples

When adding positive and negative numbers, use the following rules:

A. The sum of **two positive** numbers is **positive**.
5 + 6 = 11 Add the numbers.

For more information, see Book 6, pages 50–57.

B. The sum of **two negative** numbers is **negative**.
$-5 + (-6) = -11$ Add the numbers.

C. The sum of **a positive and a negative** number may be:

▶ **zero**

$-7.2 + 7.2 = 0$
$7.2 + (-7.2) = 0$

▶ **positive**
Find the difference between the two numbers. Take the sign of the larger number.

$-5 + 6 = 1$ Since 6 is greater than 5,
$6 + (-5) = 1$ the answer is positive.

▶ **negative**
Find the difference between the two numbers. Take the sign of the larger number.

$7 + (-9) = -2$ Since 9 is greater than 7,
$-9 + 7 = -2$ the answer is negative.

To **subtract** a number, **add** its **opposite**.

D. $\begin{array}{r} 7 \\ \underline{-(9)} \end{array} \rightarrow \begin{array}{r} 7 \\ \underline{+(-9)} \\ -2 \end{array}$ The opposite of 9 is −9.

E. $\begin{array}{r} 9 \\ \underline{-(7)} \end{array} \rightarrow \begin{array}{r} 9 \\ \underline{+(-7)} \\ 2 \end{array}$ The opposite of 7 is −7.

F. $\begin{array}{r} -7 \\ \underline{-(9)} \end{array} \rightarrow \begin{array}{r} -7 \\ \underline{+(-9)} \\ -16 \end{array}$ The opposite of 9 is −9.

G. $\begin{array}{r} -9 \\ \underline{-(7)} \end{array} \rightarrow \begin{array}{r} -9 \\ \underline{+(-7)} \\ -16 \end{array}$ The opposite of 7 is −7.

H. $\begin{array}{r} 7 \\ \underline{-(-9)} \end{array} \rightarrow \begin{array}{r} 7 \\ \underline{+9} \\ 16 \end{array}$ The opposite of −9 is 9.

I. $\begin{array}{r} 9 \\ \underline{-(-7)} \end{array} \rightarrow \begin{array}{r} 9 \\ \underline{+7} \\ 16 \end{array}$ The opposite of −7 is 7.

J. $\begin{array}{r} -7 \\ \underline{-(-9)} \end{array} \rightarrow \begin{array}{r} -7 \\ \underline{+9} \\ 2 \end{array}$ The opposite of −9 is 9.

K. $\begin{array}{r} -9 \\ \underline{-(-7)} \end{array} \rightarrow \begin{array}{r} -9 \\ \underline{+7} \\ -2 \end{array}$ The opposite of −7 is 7.

Practice

Use the number line below to find the sums.

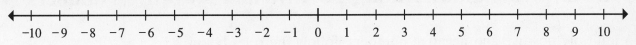

−10 −9 −8 −7 −6 −5 −4 −3 −2 −1 0 1 2 3 4 5 6 7 8 9 10

1. $3 + 1$ _____	**2.** $-3 + (-1)$ _____	**3.** $-3 + 1$ _____
4. $3 + (-1)$ _____	**5.** $2 + 5$ _____	**6.** $-2 + (-5)$ _____

Find the sums.

7. $6 + (-5)$ _____	**8.** $3 + 5$ _____	**9.** $-2 + (-4)$ _____
10. $-12 + 4$ _____	**11.** $13 + (-26)$ _____	**12.** $-8 + (-21)$ _____
13. $-54 + (-32)$ _____	**14.** $-36 + 2\frac{1}{4}$ _____	**15.** $44.5 + (-15)$ _____
16. $-26 + 15.9$ _____	**17.** $-23 + (-8)$ _____	**18.** $32 + (-32)$ _____

Subtract.

19. $-8 - 7$ _____	**20.** $-9 - 2$ _____	**21.** $-14 - (-3)$ _____
22. $6 - (-4)$ _____	**23.** $12 - (-5)$ _____	**24.** $13 - \left(-7\frac{3}{4}\right)$ _____
25. $31 - 22$ _____	**26.** $-65 - (-82)$ _____	**27.** $-41 - 55.6$ _____
28. $28 - 37$ _____	**29.** $148 - (-148)$ _____	**30.** $-65 - (-65)$ _____

Problem Solving

Solve the following problems.

31. The average temperature for bath water is 110° F. Fresh food is stored at 45° F. What is the difference in temperature?

____

32. The temperature inside a commercial freezer is approximately −5° F. Fresh food is stored at 45° F. What is the difference in temperature between the freezer and fresh food storage?

____

Multiplying and Dividing Positive and Negative Numbers

To multiply positive and negative numbers, follow the rules below.

1. The product of **two positive** numbers or **two negative** numbers is **positive.**
2. The product of a **positive** and a **negative** number is **negative.**

Examples

A. -12×-4
You are multiplying two negative numbers, so the product is positive.

$$(-12)(-4) = 48$$

B. 5×-8
You are multiplying a positive and a negative number, so the product is negative.

$$5(-8) = -40$$

C. $5 \times -2 \times -3$

$$\underbrace{(5) \times (-2)}_{(-10)} \times (-3)$$

$$(-10) \times (-3)$$

$$30$$

MATH HINT

The product of an even number of negative numbers is positive. The product of an odd number of negative numbers is negative.

To divide positive and negative numbers, use the same sign rules you used for multiplication.

D. $18 \div -6$
The quotient of a positive number divided by a negative number is negative.

$$\frac{18}{-6} = -3$$

 For more information, see Book 6, pages 60–62; 64–65.

E. $-20 \div -5$

The quotient of a negative number divided by a negative number is positive.

$$\frac{-20}{-5} = 4$$

F. $15 \div 3$

The quotient of two positive numbers is positive.

$$\frac{15}{3} = 5$$

G. $-12 \div -4$

The quotient of two negative numbers is positive.

$$\frac{-12}{-4} = 3$$

MATH HINT

Division by zero is undefined. $\frac{7}{0}$ is **undefined**. Division of zero by any number is zero. $\frac{0}{7} = 0$

Practice

Multiply or divide.

1. $7(-3)$ _____ **2.** $-5(6)$ _____ **3.** $-8(-7)$ _____

4. $(-4)(-9)$ _____ **5.** $-9(9)$ _____ **6.** $12(-4)$ _____

7. $(13)(-5)(-3)$ _____ **8.** $\left(-\frac{1}{4}\right)(-5)$ _____ **9.** $\left(\frac{2}{3}\right)(-3)$ _____

10. $1 \div -8$ _____ **11.** $6 \div \frac{1}{6}$ _____ **12.** $-235\left(-\frac{1}{235}\right)$ _____

13. $48\left(\frac{-1}{48}\right)$ _____ **14.** $12(-3)$ _____ **15.** $65 \div (-5)$ _____

16. $-39 \div (13)$ _____ **17.** $-26 \div 2$ _____ **18.** $-36 \div (-18)$ _____

19. $(-6)(-3)(10)(-8)$

____

20. $(-7)(-7)(-7)(-3)(-4)(-5)$

____

21. $(-15)(-28)(-11)(0)$

____

22. $(-100)(0)(89)(-2)$

____

23. $(5)(11)(-2)(-15)$

____

24. $(3 + 2) \div (4 - 6)$

____

25. $-80 \div 0$

____

26. $0 \div 28$

____

LESSON 57

Order of Operations

Follow these steps for solving problems with several operations:

Step 1 Do all work inside parentheses and any work above and below fraction bars.

Step 2 Simplify exponents and roots.

Step 3 Start at the left and do multiplication and division working toward the right.

Step 4 Start at the left and do addition and subtraction working toward the right.

MATH HINT

The first letters of "**P**lease **E**xcuse **M**y **D**ear **A**unt **S**ally" will help you remember the correct order.

P	**P**arentheses and fraction bars
E	**E**xponents and roots
MD	**M**ultiplication and **D**ivision
AS	**A**ddition and **S**ubtraction

Examples

A.

$6 - 4(-3 + 2) + 3^2 - 12$	parentheses
$6 - 4(-1) + 3^2 - 12$	exponents
$6 - 4(-1) + 9 - 12$	multiplication
$6 - (-4) + 9 - 12$	change subtraction to addition
$6 + 4 + 9 + (-12)$	addition
$19 + (-12)$	addition
7	

B.

$\frac{-2(6 - 3)}{(-3)^2 - 4}$	parentheses (numerator)
$\frac{-2(3)}{(-3)^2 - 4}$	multiply (numerator)
$\frac{-6}{(-3)^2 - 4}$	exponent (denominator)
$\frac{-6}{9 - 4}$	subtract (denominator)
$\frac{-6}{5}$, or $-1\frac{1}{5}$	

For more information, see Book 6, pages 68–69.

Practice

Solve.

1. $7(-5) + 3^2$ ______

2. $4^2 + 3^2 + 6(-4 + 3)$ ______

3. $9 - 6(-3)$ ______

4. $11^2 + 4(3^2 - 10)$ ______

5. $15 \div 3(-2)$ ______

6. $24 + 20 \div 2$ ______

7. $6 \div 3 \times 2 \div 2$ ______

8. $8 + 4(2 - 5) + 4^2 - 10$ ______

9. $\frac{-30}{6(5 - 2^2)}$ ______

10. $\frac{5^2 + 5}{5 + 5^2}$ ______

11. $\frac{5(-3 + 2)}{-7(5)}$ ______

12. $\left(\frac{-1}{2}\right)^2 + \frac{3}{4}$ ______

13. $2(-3)^3 - 4(-4 + 1)$ ______

14. $7^3 - (7^2 - 7)$ ______

15. $\left(\frac{2}{3}\right)^2 \div \left(\frac{3}{2}\right)^2$ ______

16. $(0.3)^2 + 1.05 - 2.1$ ______

17. $-0.5(1.2) - (0.2)^2$ ______

18. $7(3) \div 3 + 3(-7) \div 7$ ______

19. $\frac{(0.1)^2 - 0.01}{-(0.1)}$ ______

20. $\frac{\frac{3}{5} + \left(\frac{-1}{5}\right)^2}{\frac{-1}{5}}$ ______

Problem Solving—Classifying a Problem

The steps you have learned to help solve word problems can be used with word problems that deal with positive and negative numbers. Use the following steps:

Step 1 Read the problem and underline the key words. These words will usually relate to some mathematical reasoning.

Step 2 Make a plan to solve the problem. Ask yourself, Should I add, subtract, multiply, divide, round, or compare? You may have to do more than one of these operations for the same problem.

Step 3 Find the solution. Use your math knowledge to find your answer.

Step 4 Check your answer. Ask yourself, Is the answer reasonable? Did you find what you were asked for?

Example

Kelly's credit card balance is $75. He uses the card for purchasing merchandise at several stores and spends a total of $260. He then pays the minimum payment of $35. What is the new, or ending, balance for the credit card?

Step 1 Read the problem and identify the key words.
The key words are **spend**, **pay**, and **total**.

Step 2 Make a plan to solve the problem.
The problem can be classified as an addition of signed numbers problem. The amounts Kelly owes are negative numbers. The amounts owed are subtracted from the amount of credit balance he can have on the credit card. The payments he makes are positive numbers. The amounts paid are added to the credit card balance.

Step 3 Find the solution.

beginning credit card balance	−$75.00
purchases	−$260.00
payment	+35.00
ending balance	−$300.00

Kelly owes $300.

For more information, see Book 6, pages 70–72.

Step 4 Check your answer.

$$\begin{array}{r} -\$300.00 \\ -\$35.00 \\ +\$260.00 \\ \hline -\$75.00 \end{array}$$

The problem checks working backwards.

Practice

Solve the following problems.

1. The value of a share of stock rose $2.35 and then dropped $3.25 before finishing the day at $64.35. What was the original value of the stock per share?

____

2. The value of a baseball card dropped $12 and then rose $3.70 before settling at a value of $32.50. What was its original value?

____

3. In a college football game, the quarterback attempted four passes with the results as shown in the chart. Find the total gain or loss for passing.

Try	Gain or Loss
1	10-yard gain
2	0-yard gain
3	21-yard loss
4	14-yard gain

____

4. The table below shows the profit and loss of a small business over a five-year period. Find the profit or loss after this period of time.

Year	Profit or Loss
1990	+$32,000
1991	−$15,000
1992	+$25,000
1993	−$10,000
1994	−$5,000

____

5. After five rounds of golf, a golfer was three under par twice, two over par once, two under par once, and one under par once. At the end of the day, how far above or below par was the golfer?

____

6. In Canada, the average low temperature in January is -31°C. In Florida, the average low temperature in January is 19°C. What is the difference in the average low temperatures of the two cities?

____

7. On one winter day, the temperature dropped from 45°F to -15°F. How many degrees did the temperature drop?

____

8. A savings account had the following transactions during the course of one month.

Deposit	$100
Withdrawal	$50
Deposit	$140
Deposit	$175
Withdrawal	$75
Withdrawal	$120

What was the balance at the end of the month?

____

LIFE SKILL

Balancing a Checking Account

When you write checks or deposit money into your checking account, you need to record the transaction. Positive and negative numbers can help you to keep track of the money in your account.

When you receive your statement from the bank each month, it is good to balance your account to make sure the account is accurate.

To balance your checking account, follow these steps:

Step 1 Write down the balance that is shown on the statement you receive from the bank.

Step 2 Add the amount of the deposits you have made that do not appear on the statement.

Step 3 Subtract the total amount of checks that you have written that do not appear on your statement. The new total is the adjusted statement total.

Step 4 Write down the balance from your checkbook.

Step 5 Add any deposits that are on the statement but not in your checkbook.

Step 6 Subtract all charges on your statement that you have not already subtracted in your checkbook. The new total is the adjusted checkbook balance.

Step 7 Compare the adjusted totals. They should be equal if the account balances.

Use the following information to balance the checking account below.

Information on bank statement		Information in checkbook	
Balance	$2,973.24	Balance	$2,891.30
Service charge	$2.51	Deposits	$243, $182
Automatic deposit	$279.95	Checks outstanding	$140.50
			$12.75
			$76.25

Balance shown on bank statement ______________

Add deposits not on statement ______________

____

Subtotal ______________

Subtract outstanding checks ______________

____

____

Balance ______________

Balance shown in checkbook ______________

Add deposits not entered in checkbook ______________

Subtotal ______________

Subtract bank charges ______________

Balance ______________

Does the checking account balance? ______________

UNIT 9

Posttest

Place the correct symbol, $<$, $>$, or $=$ between the two numbers to make a true statement.

1. 7 _____ -8

2. -1 _____ 0.3

3. -7 _____ 5

4. $-(-4)$ _____ 2

5. $11(-3)$ _____ $3(11)$

6. $-(-1)$ _____ -2

Add or subtract.

7. $6 - (-4)$ _____

8. $1 + (-26)$ _____

9. $(-21) - 44$ _____

10. $32 + (-18.2)$ _____

11. $3 - (-3)$ _____

12. $-6 + (-5)$ _____

13. $36 - 75$ _____

14. $28 + -57$ _____

15. $-18 - (-15\frac{4}{5})$ _____

Multiply or divide.

16. $-18(-7)$ _____

17. $(3)(-21)$ _____

18. $(-3)(14)(-7)$ _____

19. $\frac{6}{(-2)}$ _____

20. $35\left(-\frac{1}{5}\right)$ _____

21. $\frac{-2}{(48)}$ _____

22. $(-6)(12)(2)$ _____

23. $(-5)(-20)(-2)$ _____

24. $[(5) + (11)][2 - (15)]$ _____

25. $\frac{-120}{0}$ _____

26. $5 + 3^2 - 2(4 - 5)$ _____

27. $\frac{+5(-1 - (3)^2)}{3(5^2)}$ _____

Problem Solving

Solve the following problem.

28. Tony has a checking account with a balance of \$540. He writes checks for \$150 and \$125. He deposits \$315 into the account and writes another check for \$340. What is the new balance?

UNIT 10

Algebraic Expressions and Equations

Pretest

Write an algebraic expression for each phrase.

1. A number plus 6 ____________

2. 25 less than a number ____________

3. 5 times the quotient of a number and 10 ____________

4. 25 less than one more than a number ____________

Evaluate the following.

5. $d = rt$
If $r = 50$ and $t = 2$, find d.

____

6. $S = P - 0.25P$
If $S = \$30$, find P.

____

7. $P = 2l + 2w$
If $l = 8$ feet and $w = 4$ feet, find P.

____

8. $V = lwh$
If $V = 45$ cubic feet, $l = 5$ feet, and $h = 3$ feet, find w.

____

Combine like terms.

9. $8a + 11a - 9$ __________

10. $10ac + 6ac + 4ab - 3ab$ __________

11. $4x - 11y - 8x$ __________

12. $-3v^2 - 5v - 2v^2 + 7v$ __________

Solve these equations for x.

13. $2(2x + 1) = 12$ __________

14. $-4(3x + 2) = 28$ __________

15. $5x + 20 = 15$ __________

16. $22 = 8x + 18$ __________

17. $x + 0.2 = 2.8$ __________

18. $0.5x - 3.5 = 2$ __________

19. $3x - 24 + x = 12$ __________

20. $7x - 5x + 3 = 31$ __________

21. $-(x + 5) - 3x = 11$ __________

22. $2x + 2(x - 6) = 10$ __________

Problem Solving

Solve the following problems.

23. A student has an average score on tests of 82. So far, the student has taken 3 tests. Two of the scores are 75 and 80. What is the missing test score?

____

24. Calvin's take-home pay is $2,100 each month. He pays $600 per month for house expenses. If he spends 20% of his remaining income on food, how much does he spend on food each month?

____

Algebraic Expressions

In algebra, you use letters to represent unknown numbers. When using algebra to solve problems, you have to change words into symbols. Study the following lists to see the most common key words used to indicate the four basic arithmetic operations.

Addition	**Subtraction**	**Multiplication**	**Division**	**Equals**
sum	less than	times	divided by	is
plus	minus	product	quotient	equals
more than	difference	of	ratio	results
increased by	taken from	multiplied by	divided into	same as
added to		squared		

An **algebraic expression** is made up of numbers, letters, and operation signs.

Examples

A. Write an algebraic expression for each of the following word statements.

Word Statement	**Algebraic Expression**
A number increased by two	$x + 2$
Nine subtracted from a number	$x - 9$
The product of seven and a number	$7x$
A number divided by twelve	$\frac{x}{12}$
Twelve divided by a number	$\frac{12}{x}$
Four-fifths of a number	$\frac{4x}{5}$ or $\frac{4}{5}x$

MATH HINT

A number and a letter together indicate multiplication.

B. Notice in the following examples that when *less than* or *more than* are used, the number mentioned first is written second.

Eight less than a number	$x - 8$
A number less than 8	$8 - x$
Eight more than a number	$x + 8$
A number more than 8	$8 + x$

For more information, see Book 6, pages 77–79.

Practice

Write an algebraic expression for each phrase.

1. A number plus 2

   ____

2. 6 times a number

   ____

3. A number less 10

   ____

4. A number divided by 5

   ____

5. The square of a number

   ____

6. 2 divided into a number

   ____

7. A number increased by 4

   ____

8. The sum of a number and -7

   ____

9. A number less 12

   ____

10. 12 less than a number

    ____

11. 10 more than twice a number

    ____

12. 4 times the quotient of a number and 3

    ____

13. A number less 15% of the number

    ____

14. Five more than the square of a number

    ____

15. One fifth of one more than a number

    ____

16. The age of a 50-year-old person x years ago

    ____

Evaluating Algebraic Expressions

The value of an algebraic expression depends upon the numbers which replace the variables.

To evaluate algebraic expressions, follow these steps:

Step 1 Replace each variable with its value (if known).

Step 2 Simplify using the **order of operations** and the **rules for signed numbers.**

Example

A. Evaluate $2x^2 - (3 + y)$ if $x = -2$ and $y = 5$.

$$2(-2)^2 - (3 + 5)$$
$$2(4) - 8$$
$$8 - 8$$
$$0$$

So, $2x^2 - (3 + y) = 0$ when $x = -2$ and $y = 5$.

A formula is an equation that tells us how to calculate a special problem. You have used formulas in the geometry units. The formula for calculating the area of a rectangle is $A = lw$, where l stands for the length of the rectangle and w stands for the width of the rectangle. The formula tells you to multiply l and w to calculate the area.

Example

B. The area (A) of a square is equal to s (the length of the side) squared. Evaluate the formula if $s = 1.2$ cm.

$$A = s^2$$
$$A = (1.2)^2$$
$$A = 1.44 \text{ cm}^2$$

For more information, see Book 6, pages 84–87.

Practice

Evaluate each expression for the given values.

1. $A = lw$
   $l = 10$, $w = 5.2$ __________

2. $P = 2l + 2w$
   $l = 10$, $w = 3$ __________

3. $A = \frac{1}{2}bh$
   $b = 4$, $h = 5$ __________

4. $F = 1.8C + 32$
   $C = 50$ __________

5. $C = \frac{5}{9}(F - 32)$
   $F = 95$ __________

6. $I = prt$
   $p = 2{,}000$, $r = 0.075$
   $t = 2$ __________

7. $p = c - 0.05c$
   $c = 40$ __________

8. $d = rt$
   $r = 55$, $t = 3.5$ __________

Write the formula described in each of the following problems. Then evaluate the formula using the indicated values.

9. The area (A) of a square is found by squaring the length of its side (s). Solve for A if $s = 6.25$ inches.

   ____

10. The perimeter (P) of a square is found by multiplying the length of its side (s) by 4. Solve for P if $s = 6.25$ inches.

    ____

11. The area (A) of a rectangle is found by multiplying its length (l) by its width (w). Solve for A if $l = 7\frac{1}{4}$ yd and $w = 4\frac{3}{4}$ yd.

    ____

12. The perimeter (P) of a rectangle is the sum of twice its length (l) and twice its width (w). Solve for P if $l = 7\frac{1}{4}$ yd and $w = 4\frac{3}{4}$ yd.

    ____

13. To find the average (M) of 4 test grades (A, B, C, and D), divide their sum by 4. Solve for M if $A = 70$, $B = 95$, $C = 92$, and $D = 43$.

    ____

14. To find simple interest (I), multiply the principal (p) by the annual rate (r) by the time in years (t). Solve for I if $p = \$640$, $r = 9\%$, and $t = 2.5$ years.

    ____

15. To find the volume (V) of a rectangular box, multiply the length (l) by the width (w) and the height (h). Solve for V if $l = 8$ inches, $w = 4.2$ inches, and $h = 3\frac{1}{4}$ inches.

____

16. The time (t) needed to travel a distance (d) is found by dividing the distance (d) by the average rate (r). Solve for t if $d = 275$ miles and $r = 55$ mph.

____

17. The total cost (c) of an item is its price (p) plus the sales tax, which is the sales tax rate (r) times the price (p). Solve for c if $p = \$25$ and $r = 5.75\%$.

____

18. The sale price (s) of an item on sale is the regular price (p) minus the product of the regular price (p) and the discount rate (d). Solve for s if $p = \$95$ and $d = 15\%$.

____

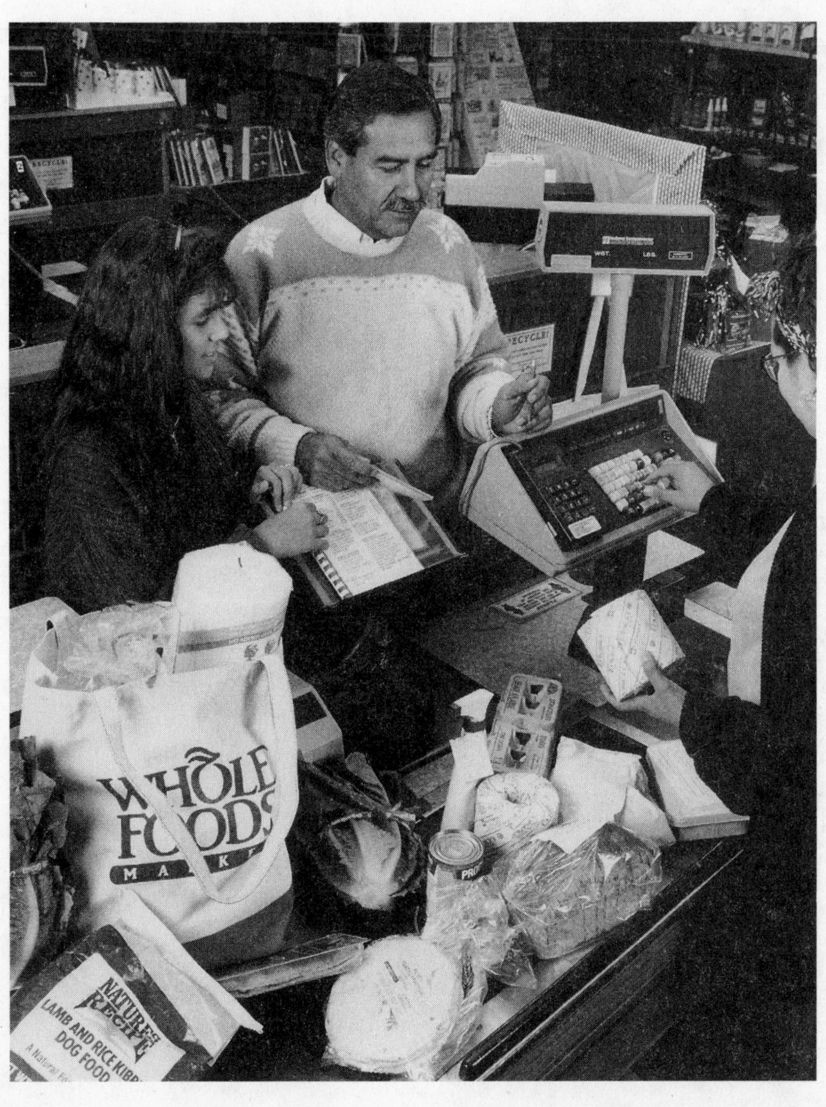

LESSON 61

Combining Like Terms

A **term** is a number, letter, or numbers and letters combined by multiplication, division, or both. A numerical coefficient is the number in front of the term. If there is no number, it is understood to be 1.

$4x^2 + 3x^2 + 4x + 5x - 8 + 3$ is an algebraic expression that has 6 terms.
4 is the numerical coefficient for $4x^2$, 3 is the numerical coefficient of $3x^2$, 4 is the numerical coefficient of $4x$, and 5 is the numerical coefficient of $5x$.

Like terms have the same variables with the same exponents.

Pairs of like terms: $4x^2$, $3x^2$; $4x$, $5x$; -8, 3

Unlike terms differ either in their variables and/or in their exponents.

$4x$ and $4x^2$ are unlike terms.

Like terms are combined by adding or subtracting their numerical coefficients. Simplify the following algebraic expression:

$$4x^2 + 3x^2 + 4x + 5x - 8 + 3$$

$$\underbrace{4x^2 + 3x^2}_{7x^2} + \underbrace{4x + 5x}_{+9x} - \underbrace{8 + 3}_{-5}$$

Examples

A. Simplify $-4y + 8x + 9y$.
$-4y$ and $9y$ are like terms.
Combining, you get $5y + 8x$.
You cannot combine $5y$ and $8x$ because the letters, or variables, are different.

 For more information, see Book 6, pages 81–83.

B. Simplify $-4z + 2z^2 - 6z^2 - 5z + 7x + x$.

Like terms	**Combining**
$-4z, -5z$	$-9z$
$2z^2, -6z^2$	$-4z^2$
$7x, x$	$8x$

The expression is $8x - 9z - 4z^2$.

To eliminate parentheses, multiply everything inside the parentheses by the number directly in front of them.

C. Simplify $5(x + 3)$.

$$\begin{aligned} 5(x + 3) &= 5(x) + 5(3) \\ &= 5x + 15 \end{aligned}$$

D. Simplify $7x(x - 3) + 6x$.

$$\begin{aligned} 7x(x - 3) + 6x &= 7x(x) - 7x(3) + 6x \\ &= 7x^2 - 21x + 6x \\ &= 7x^2 - 15x \end{aligned}$$

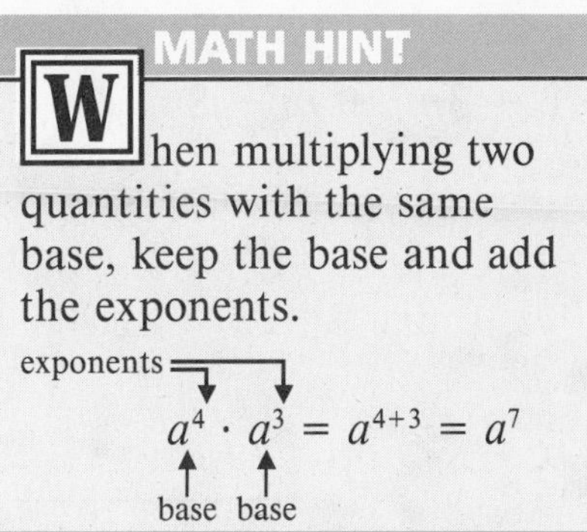

MATH HINT
When multiplying two quantities with the same base, keep the base and add the exponents.
exponents
$a^4 \cdot a^3 = a^{4+3} = a^7$
base base

E. Simplify $x + 3(x + 2)$.

$$\begin{aligned} x + 3(x + 2) &= x + 3x + 6 \\ &= 4x + 6 \end{aligned}$$

F. Jason stops every day—Monday through Friday—on his way to work to buy a cup of coffee for 60 cents and a croissant for \$1.10. How much does he spend each week?

To find the total amount per week, multiply

$$\begin{aligned} 5(0.60 + 1.10) &= 5(0.60) + 5(1.10) \\ &= 3.00 + 5.50 \\ &= 8.50 \end{aligned}$$

Jason spends \$8.50 per week.

Practice

Combine like terms to simplify the expressions.

1. $5x + 2x$ ________
2. $-3y + 5y$ ________
3. $-5a + 3a$ ________
4. $-10abc + 6abc$ ________
5. $x^2y + 3x^2y$ ________
6. $-6ab + 6ab$ ________
7. $x + 7y - 2x$ ________
8. $5x - 6y + x$ ________
9. $3x^2 + 12x - 4x^2 + 5x$ ________
10. $-13m^2 - 15m - m^2 - 7m$ ________
11. $m^2 + 3m + 2 - 3m^2 + m - 8$ ________
12. $4z + 2z^2 - 6z^2 + 4z + 4z^2$ ________

Simplify.

13. $2(x + 5)$ ________
14. $3(2x + y)$ ________
15. $4(x + 2y)$ ________
16. $4(x - 3)$ ________
17. $-4(x - 5y)$ ________
18. $-3(3x - 2y)$ ________
19. $3x + 2(x + 1)$ ________
20. $2x + 3(2x - 1)$ ________
21. $x + 5(x + 4)$ ________
22. $4(2x + y) - 6(4x - 3y)$ ________

Problem Solving

Write an algebraic expression and solve the following problems.

23. Heidi earns \$540 a week. Each week \$190 is deducted for taxes, insurance, and savings. How much is Heidi's take-home pay during a 4-week period?

____

24. Toby earns a monthly salary of \$1,250 and a monthly commission of \$900. What are his yearly earnings?

____

Solving Equations

An **equation** is a statement that two quantities are equal. To solve an equation, get the variable, or unknown number, on one side of the equal sign and its value on the other side. To do this, use the opposite operation of the one shown in an equation. Any operation performed on one side of an equation must also be performed on the other.

The opposite operation of *addition* is *subtraction.*
The opposite operation of *subtraction* is *addition.*
The opposite operation of *multiplication* is *division.*
The opposite operation of *division* is *multiplication.*

Examples

A. Solve the equation $x - 5 = 11$.

$$x - 5 = 11$$

Subtraction is indicated. Add 5 to both sides of the equation to find the answer.

$$\begin{aligned} x - 5 &= 11 \\ +5 &= +5 \\ \hline x &= 16 \end{aligned}$$

Check:
$$\begin{aligned} x - 5 &= 11 \\ 16 - 5 &= 11 \\ 11 &= 11 \checkmark \end{aligned}$$

B. Solve the equation $x + 7 = 19$.

$$x + 7 = 19$$

Addition is indicated. Subtract 7 from both sides of the equation to find the answer.

$$\begin{aligned} x + 7 &= 19 \\ -7 &= -7 \\ \hline x &= 12 \end{aligned}$$

Check:
$$\begin{aligned} x + 7 &= 19 \\ 12 + 7 &= 19 \\ 19 &= 19 \checkmark \end{aligned}$$

For more information, see Book 6, pages 98–103.

C. Solve the equation $4x = 20$.

$$4x = 20$$
$$\frac{4x}{4} = \frac{20}{4}$$
$$x = 5$$

Multiplication is indicated. Divide both sides of the equation by 4 to find the solution.

Check: $4x = 20$
$$4(5) = 20$$
$$20 = 20 \checkmark$$

D. Solve the equation $\frac{1}{2}x = 9$.

$$\frac{1}{2}x = 9$$
$$\frac{\frac{1}{2}x}{\frac{1}{2}} = \frac{9}{\frac{1}{2}}$$
$$2\left(\frac{1}{2}x\right) = 2(9)$$
$$x = 18$$

Multiplication is indicated. Divide both sides of the equation by $\frac{1}{2}$. To divide by $\frac{1}{2}$, multiply by the reciprocal of $\frac{1}{2}$, which is 2.

Check: $\frac{1}{2}x = 9$
$$\frac{1}{2}(18) = 9$$
$$9 = 9 \checkmark$$

Practice

Solve these equations. Check by substituting your answer in the original equation.

1. $5x = 20$ __________

2. $45 = 9x$ __________

3. $\frac{1}{3}x = 6$ __________

4. $10 = \frac{1}{4}x$ __________

5. $x + 9 = 15$ __________

6. $18 = x - 5$ __________

7. $x - 2 = 4.2$ __________

8. $x + 2 = 4.2$ __________

9. $2x = 4.2$ __________

10. $\frac{x}{2} = 4.2$ __________

11. $x - 8 = 12$ __________

12. $20 = x - 9$ __________

13. $32 = x + 18$ __________

14. $x - 5 = -7$ __________

15. $x + 7 = 3$ __________

16. $5 = x + 12$ __________

17. $5 + x = -2$ __________

18. $-8 = -7 + x$ __________

19. $-5 = x - 5$ __________

20. $x + 8 = 8$ __________

21. $x + 0.5 = 0.9$ __________

22. $x - 1.2 = 0.8$ __________

23. $0.7 + x = 0.3$ __________

24. $-0.2 = x + 0.4$ __________

25. $x + 3.5 = 2$ __________

26. $4x = -12$ __________

LIFE SKILL

Open-End Credit

Some credit card companies offer accounts that are open-ended. This means you make regular payments and can still borrow at a stated interest rate.

There are two main methods to figure interest for these accounts. Some companies charge interest on the unpaid balance at the end of each month. Other companies charge interest on the average daily balance.

Unpaid Balance Method

Rob has a charge card with an unpaid balance. At the beginning of June, the unpaid balance was \$150. During the month, he made a payment of \$42.50. He charged \$356 worth of goods during the month. The bank charged a finance charge of 1.7% per month on the unpaid balance. What was Rob's new unpaid balance at the end of the month?

First, find the interest on the unpaid balance.

$$\$150 \times 0.017 = \$2.55$$

The interest on the unpaid balance is \$2.55.

To find the new unpaid balance at the end of the month, first add the finance charge and the purchases to the beginning unpaid balance. Then subtract any payments made during the month.

unpaid balance		finance charge		purchases in month		payment		new unpaid balance
↓		↓		↓		↓		↓
\$150	+	\$2.55	+	\$356	−	\$42.50	=	\$466.05

Average Daily Balance Method

Karen has a credit card that charges interest using the average daily balance method. Her account has the following activity for the month of November.

Date	Activity	
November 1	Previous balance	$100
November 15	Charged	$150
November 20	Paid	$60
November 28	Charged	$50

To calculate her interest, the company needs to figure out Karen's average daily balance. To do this, add the balance for each day and divide by the number of days in the month.

For 15 days (November 1 to November 15), her balance was $100.
For 5 days (November 16 to November 20), her balance was $250.
For 8 days (November 21 to November 28), her balance was $190.
For 2 days (November 29 and November 30), her balance was $240.

$$\begin{aligned}\text{Average daily balance} &= \frac{15(100) + 5(250) + 8(190) + 2(240)}{30} \\ &= \frac{1{,}500 + 1{,}250 + 1{,}520 + 480}{30} \\ &= \frac{4{,}750}{30} \\ &= \$158.33\end{aligned}$$

The average daily balance is then multiplied by the interest being charged. Karen's bank charges 1.5% per month on the average daily balance.

$$\$158.33 \times 0.015 = 2.37495$$

The interest charge for the month of November is $2.37. Karen's new unpaid balance is $240 + $2.37, or $242.37.

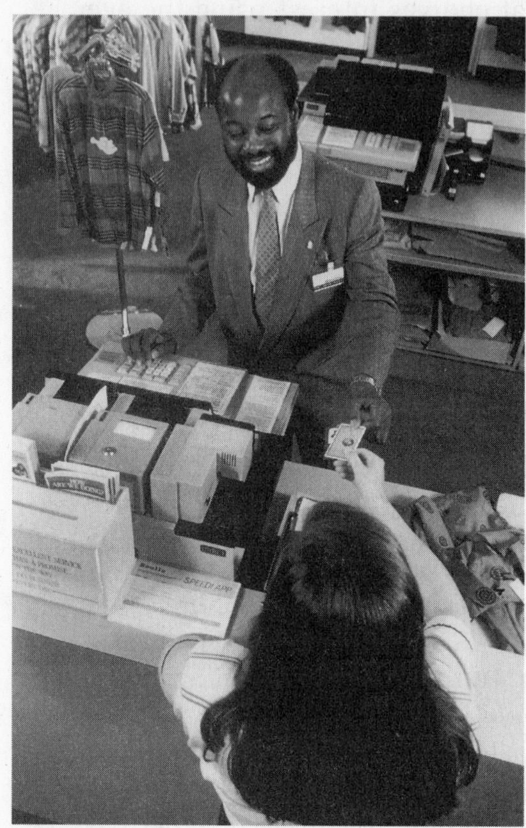

Compare the interest paid on each account by answering the following questions.

1. Find the interest on Karen's account using the unpaid balance method. ______________

2. Rob's account activity was as follows. Determine his interest using the average daily balance method. ______________

Date	Activity	
June 1	Previous balance	$150
June 10	Charges	$356
June 20	Paid	$42.50

Solving Equations With More Than One Step

Sometimes you must do more than one step to solve an equation. When solving such equations, follow these steps:

Step 1 Multiply to eliminate parentheses.

Step 2 Combine like terms to simplify both sides as much as possible.

Step 3 Use addition and subtraction to move the terms with the variable x in them to one side of the equation and the numbers to the other side.

Step 4 Use multiplication or division to find the value of the variable.

Step 5 Check the solution by substituting the answer into the original equation.

MATH HINT

When solving equations, you can remove fractions by multiplying each term by the lowest common denominator. You can remove decimals by multiplying each term by a power of 10 large enough to make all decimals become whole numbers.

Examples

A. Solve $10 + 4x + 20 = -2(3x - 5)$.

Step 1	Remove the parentheses.	$10 + 4x + 20 = -6x + 10$
Step 2	Combine like terms.	$30 + 4x = -6x + 10$
Step 3	Subtract 30 from both sides of the equation.	$30 + 4x - 30 = -6x + 10 - 30$ $4x = -6x - 20$
	Add $6x$ to both sides of the equation.	$4x + 6x = -6x - 20 + 6x$ $10x = -20$
Step 4	Multiply by the reciprocal of $\frac{1}{10}$.	$\frac{1}{10}(10x) = \frac{1}{10}(-20)$ $x = -2$
Step 5	Check.	$10 + 4x + 20 = -2(3x - 5)$ $10 + 4(-2) + 20 = -2[3(-2) - 5]$ $10 - 8 + 20 = -2(-6 - 5)$ $22 = -2(-11)$ $22 = 22\checkmark$

For more information, see Book 6, pages 106–109.

Practice

Solve these equations. Then check your answers.

1. $3x + x = 20$ ________

2. $9x - x = 56$ ________

3. $5x + 7x = 144$ ________

4. $\frac{2}{5}x + \frac{1}{5}x = 15$ ________

5. $\frac{1}{4}x + \frac{1}{2}x = 12$ ________

6. $2x + 3 = 11$ ________

7. $4x - 5 = 11$ ________

8. $55 = 20x - 5$ ________

9. $31 = 2x + 1$ ________

10. $4x - 2 = -34$ ________

11. $3x - 5 = -23$ ________

12. $8 - 3x = -4$ ________

13. $x - 8 + 2x = 4$ ________

14. $4x + 15 - x = 12$ ________

15. $9x - 5 + 3x = 31$ ________

16. $5x - 23 + x = 7$ ________

17. $2x + 25 - 5x = 1$ ________

18. $90 = 8x + 15 - 3x$ ________

19. $x + (x - 5) = 21$ ________

20. $x + (2x - 3) = 24$ ________

21. $2x - (x + 6) = 13$ ________

22. $3x - (x + 5) = 25$ ________

23. $x - (6 - x) = 30$ ________

24. $2x - (5 - x) = 25$ ________

25. $5x - (2x + 9) = 0$ ________

26. $8x - (3x + 15) = 0$ ________

27. $0.2x - (0.1x - 5) = 0$

____

28. $0.3x - (0.2x + 3) = 0$

____

29. $-0.5x + (0.3x + 0.8) = 0$

____

30. $2(x - 5) = 3(2x + 10)$

____

31. $6(x + 2) = 5(x + 3)$

____

32. $4(3x - 2) + 5 = 57$

____

33. $5(2x + 3) + 7 = 62$

____

34. $-5(2x + 3) - 3x = 11$

____

35. $12x - 2(x + 6) = 18$

____

36. $3x - (2x - 5) = 12 - 2(x + 5)$

____

37. $2(x - 3) + 7 = 13 - 4(x + 6)$

____

38. $-5(x + 3) + 7(x - 1) = 3(x - 2) - 2$

____

Problem Solving—Using Formulas

The steps you have learned to help solve word problems can be used with word problems that deal with formulas. Use the following steps:

Step 1 Read the problem and underline the key words. These words will generally relate to some mathematical reasoning.

Step 2 Make a plan to solve the problem. Ask yourself, Should I add, subtract, multiply, divide, round, or compare? You may have to do more than one of these operations for the same problem.

Step 3 Find the solution.

Step 4 Check your answer. Ask yourself, Is the answer reasonable? Did you find what you were asked for?

A formula is arranged to calculate one quantity. Sometimes you need to calculate one of the other quantities in the formula. To change the formula, treat all the other quantities as if they were known numbers. Then add, subtract, multiply, or divide to get the variable by itself.

Examples

A. The simple interest formula is $I = prt$, where I is the interest, p is the principal, r is the rate, and t is the time. Find the principal (p) to invest at 5% interest (r) for 5 years (t) to earn $500.

$$I = prt$$

Change the formula to solve for p. Notice that multiplication is indicated.

$$\frac{I}{rt} = \frac{prt}{rt}$$

The opposite of multiplication is division. Divide both sides of the equation by rt.

$$\frac{I}{rt} = p$$

$$\frac{500}{0.05(5)} = p$$

Substitute the known values into the equations.

$$\frac{500}{0.25} = p$$

$$2{,}000 = p$$

You need to invest $2,000 to receive the $500 interest.

 For more information, see Book 6, pages 92–93; 118–119.

B. Find the width of a rectangle with a perimeter of 50 feet and a length of 15 feet.

The formula for finding the perimeter of a rectangle is $P = 2l + 2w$, where p is the perimeter, l is the length, and w is the width. Solve the formula for w.

$$P = 2l + 2w$$

$$P - 2l = 2l + 2w - 2l$$ Addition is indicated, so subtract $2l$ from both sides of the equation.

$$\frac{P - 2l}{2} = \frac{2w}{2}$$ Multiplication is indicated. Divide both sides of the equation by 2.

$$\frac{P - 2l}{2} = w$$

$$\frac{50 - 2(15)}{2} = w$$ Substitute the values.

$$\frac{50 - 30}{2} = w$$

$$\frac{20}{2} = w$$

$$10 = w$$

The width is 10 feet.

Problem Solving

Solve the following problems.

1. The perimeter of a rectangle is 40 feet. If the length is 15 feet, find the width. Use the formula $P = 2l + 2w$.

____

2. An investor receives $1,000 in interest for an investment. If the interest rate is 10% for 4 years, how much is invested? Use the formula $I = prt$.

____

3. The volume of a box is 27 cubic inches. The box is a cube and has a height of 3 inches. What is the length of each side? Use the formula $V = lwh$.

____

4. The area of a circle is 628 square feet. Find the radius of the circle. Use the formula $A = \pi r^2$.

____

5. A triangle has a perimeter of 33 feet. If the figure is an equilateral triangle, what is the measure of each side? Use the formula $P = 3s$.

____

6. The temperature outside is 20° C. What is the temperature in Fahrenheit? Use the formula $F = \left(\frac{9}{5}\right)C + 32$.

____

7. The class has an average score of 78 on the midterm. There are 5 students in the class. Four of the scores are 75, 80, 85, and 70. Find the score for the fifth person.

____

8. The area of a triangle is 40 square feet. The height is 10 feet. Find the length of the base. Use the formula $A = \frac{1}{2}bh$.

____

UNIT

10

Posttest

Write an algebraic expression for each phrase.

1. 7 more than a number __________

2. A number less 15 __________

3. The quotient of a number and 10 times the number

____

4. 20 more than four times a number

____

Solve.

5. Solve $I = prt$
If $P = \$4{,}000$, $r = 5\%$, $T = 2$ years, find I.

____

6. Find the area of a square with sides measuring 4 yards. Use the formula $A = s^2$.

____

7. Calculate the number of hours an employee worked if the total earnings are $500 and the rate of pay is $10 per hour.

____

8. Find the average grade for a student with the following grades on four tests: 75, 80, 84, and 94.

____

Combine like terms.

9. $3t + 2 - 4t$

____

10. $12mn - 4m + 6n + 2mn$

____

11. $18d - 4d + 7c - 3d$

____

12. $-2p^2 + 15p - 42p^2 + 7p$

____

Solve these equations for *x*.

13. $2x + 9 = 17$

____

14. $4 + 3(3x + 2) = 28$

____

15. $4x + 120 = 28$

____

16. $100 = 6x - 14$

____

17. $2x + 200 = 12x$

____

18. $4.5x - 3.0 = 3.0$

____

19. $3 - 24y + 1 = 12y$

____

20. $92x - 45x + 30 = -64$

____

21. $(x - 4) + x = 26$

____

22. $12x - 2(x - 4) = 66$

____

Problem Solving

Solve the following problems.

23. A salesperson receives a 5% commission on all sales over $2,000. This month the salesperson earned $200 plus $400 commission. How much did he sell?

____

24. The perimeter of a square is 100 feet. What is the measure of the sides?

____

UNIT

11

Algebra Problem Solving

Pretest

Solve the following problems.

1. Twice the sum of 4 and a number is 34. Find the number.

____

2. Find two consecutive even integers whose sum is 114.

____

3. Two boats leave the same dock at the same time, headed in the same direction. If the faster boat averages 40 mph and the slower boat averages 35 mph, how long will it take for them to be 30 miles apart?

____

4. Flint bought 18 trees for $750. Pine trees cost $45 each, and oak trees cost $25 each. How many of each type of tree did he buy?

____

5. Carol has dimes and quarters worth $15.25. She has 103 coins. How many of each type of coin does she have?

____

6. Brian is twice as old as his sister Lisa. If six years are subtracted from Brian's age and six years are added to Lisa's age, their ages will be equal. What are their ages now?

____

Using Algebra to Solve Number Problems

The most difficult step in using algebra to solve problems is changing the words into symbols and the sentences into equations.

Examples

A. Change the following sentences into equations.

The sum of a number and 3 is 15.	$x + 3 = 15$
The difference of a number and 3 is 15.	$x - 3 = 15$
The product of a number and 3 is 15.	$3x = 15$
The quotient of a number and 3 is 15.	$\frac{x}{3} = 15$

Sometimes a problem will ask you to find more than one number. Remember to use the problem-solving steps.

MATH HINT

When a problem contains more than one unknown quantity, let x equal the one that you know the least about.

B. The sum of two numbers is 36. The larger is three times the smaller. What are the numbers?

Step 1 Read the entire problem very carefully. The key words are **sum**, **larger**, and **smaller**.

Step 2 Make a plan to solve the problem.
First, multiply.

Choose a variable (letter) to represent an unknown, and write down what it represents.

Let x = smaller number
Let $3x$ = larger number
Then use addition to solve the problem.

larger number + smaller number = 36
$3x + x = 36$ is the equation.

 For more information, see Book 6, pages 130–133.

Step 3 Solve the equation.

$$
\begin{aligned}
3x + x &= 36 \\
4x &= 36 \\
x &= 9
\end{aligned}
$$

The smaller number is 9. The larger number is 3(9), or 27.

Step 4 Check the answer. Substitute the values in the equation.

$$
\begin{aligned}
3x + x &= 36 \\
3(9) + 9 &= 36 \\
36 &= 36 \checkmark
\end{aligned}
$$

C. Find four consecutive odd integers so that the sum of the second and third numbers is 32.

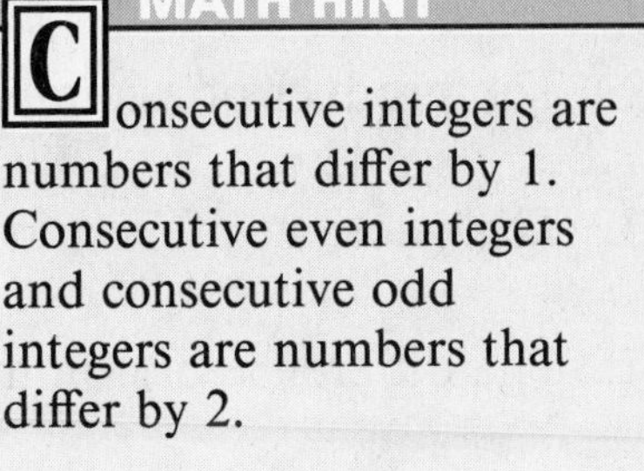
MATH HINT

Consecutive integers are numbers that differ by 1. Consecutive even integers and consecutive odd integers are numbers that differ by 2.

Step 1 Read the entire problem very carefully. The key words are **sum** and **consecutive odd integers.**

Step 2 Make a plan to solve the problem.
You need to add.
Choose a variable (letter) to represent an unknown, and write down what it represents.

Let x = first consecutive odd integer.
$x + 2$ = second consecutive odd integer
$x + 4$ = third consecutive odd integer
$x + 6$ = fourth consecutive odd integer

The expression for the problem is as follows:
third number + fourth number = 32

$$(x + 2) + (x + 4) = 32$$

Step 3 Solve the equation.

$$
\begin{aligned}
(x + 2) + (x + 4) &= 32 \\
2x + 6 &= 32 \\
2x &= 26 \\
x &= 13
\end{aligned}
$$

The first consecutive odd integer is 13. The other numbers are 15, 17, and 19.

Step 4 Check the answer. The sum of the second and third numbers should equal 32.

$$15 + 17 = 32$$
$$32 = 32 \checkmark$$

Practice

Write the following sentences as equations.

1. The sum of a number and 6 is 45. __________

2. The product of a number and 25 is 80. __________

3. The quotient of 64 and some number is 12. __________

4. The difference of a number and $\frac{1}{2}$ is 3. __________

5. Four times a number is 24. __________

Problem Solving

Solve the following problems.

6. Five more than twice a number is 47. Find the number. __________

7. Two more than three times a number is 29. Find the number. __________

8. The sum of two numbers is 25. The smaller number is 5 less than the larger number. Find the numbers. __________

9. The sum of two numbers is 30. The larger number is 6 more than the smaller number. Find the numbers. __________

10. The sum of two consecutive integers is 21. Find the numbers. __________

11. The sum of two consecutive integers is 59. Find the numbers. ________

12. Find two consecutive even integers with a sum of 38. ________

13. Find two consecutive odd integers whose sum is 44. ________

14. The sum of three consecutive odd integers is 45. Find the numbers. ________

15. The sum of three consecutive even integers is 12. Find the numbers. ________

16. Find three consecutive even integers, in which 4 times the largest number is 56. ________

17. Find three consecutive odd integers, in which the sum of the first two numbers is 24. ________

18. Find three consecutive even integers, in which the sum of the second and third is 8 more than the first. ________

19. Find three consecutive even integers, in which the sum of the second and third numbers equals 8 less than 3 times the first number. ________

Problem Solving—Using Algebra to Solve Uniform Motion Word Problems

Uniform motion problems are also called distance problems. To solve the problem, you are looking for a distance traveled at a constant speed for a specific amount of time.

Use this equation to solve uniform motion, or distance, problems:

distance = rate × time, or
$d = rt$

Remember to use the problem-solving steps as you solve distance problems.

Examples

A. How far does a train travel in 4.5 hours if its rate is 75 mph?

Step 1 Read the problem and identify the key words.
The key words are **rate (75 mph)**, time **(4.5 hours)**, and **how far**.

Step 2 Make a plan to solve the problem.
Use the distance formula, $d = rt$, where d is the distance, r is the rate, and t is the time. You know the rate and the time.

$r = 75$ mph and $t = 4.5$ hours

Step 3 Solve the equation.

$d = rt$
$d = 75(4.5)$
$d = 337.5$

The train travels 337.5 miles.

Step 4 Check your answer. Since you multiplied to solve the equation, do the opposite operation, division, to check your solution.

Does $337.5 \div 4.5 = 75$?
$75 = 75$ ✓

 For more information, see Book 6, pages 140–141.

Some distance problems involve two moving objects. The objects may be moving in the same direction or in opposite directions. You can draw a simple sketch or make a chart to help solve problems with two moving objects.

B. Two cars leave at the same time and travel toward each other from opposite ends of a 330-mile road. If one car travels 60 miles per hour and the other car travels 50 miles per hour, how long will it take for the cars to meet?

Step 1 Read the problem and identify the key words. The key words relate to the **rate of each car** and the **distance of the road**.

Rate of the first car: **60 mph.**
Rate of the second car: **50 mph.**
Distance: **330 miles.**

Draw a sketch to help you visualize the problem.

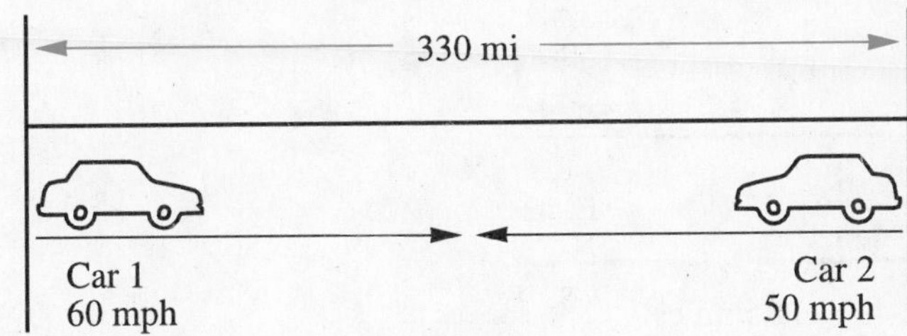

Step 2 Make a plan to solve the problem.

Set up a chart using the distance formula, $d = rt$.

	Rate	Time	Distance
car 1	50	t	$50t$
car 2	60	t	$60t$
Totals			330

The distance of the first car plus the distance of the second car equals 330.

Step 3 Find the solution.

distance for car 1 + distance for car 2 = total distance

$$\begin{aligned} 50t + 60t &= 330 \\ 110t &= 330 \\ t &= 3 \end{aligned}$$

It took the cars 3 hours to meet.

Step 4 Check your answer.

$$50t + 60t = 330$$
$$50(3) + 60(3) = 330$$
$$150 + 180 = 330 \checkmark$$

Suppose two objects travel in the same direction but at different speeds. How long will it take for the two objects to be a given distance apart? That is, how long will it take one of the objects to lead the other object by a certain distance?

C. Two cyclists leave from the same place at the same time and travel in the same direction. The faster cyclist averages 20 mph and the slower cyclist averages 15 mph. How long will it take for the cyclists to be 12.5 miles apart?

Step 1 Read the problem and identify the key words. The key words are **same direction** and **how long.** Draw a sketch to help you visualize the problem.

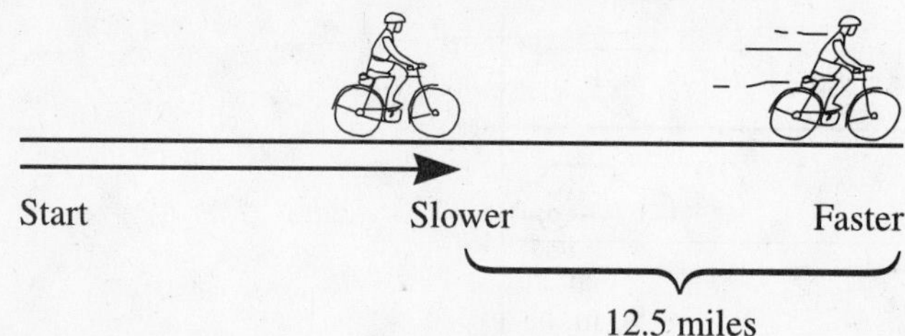

Step 2 Make a plan to solve the problem. Set up a chart using the distance formula, $d = rt$.

	Rate	Time	Distance
slower cyclist	15	t	$15t$
faster cyclist	20	t	$20t$

The distance between the cyclists is the distance of the faster cyclist minus the distance of the slower cyclist.

Step 3 Find the solution.
faster cyclist − slower cyclist = distance between

$$20t - 15t = 12.5$$
$$5t = 12.5$$
$$t = 2.5$$

It takes 2.5 hours until the distance between the cyclists is 12.5 miles.

Step 4 Check your answer.
The distance of the faster cyclist minus the distance of the slower cyclist should be 12.5 miles.

$$20(2.5) - 15(2.5) = 12.5$$
$$50 - 37.5 = 12.5$$
$$12.5 = 12.5 \checkmark$$

Practice

Solve the following problems.

1. A car travels 55 mph. How far will it go in 5 hours? __________

2. An airplane is flying at 610 mph. How far will it go in 2 hours? __________

3. A cyclist travels 108 miles at 12 mph. How long does the trip take? __________

4. Two cars start from the same point and travel in opposite directions at 50 mph and 55 mph, respectively. What is the distance between them after 4 hours? __________

5. David and Matthew each drove their classic cars from Lakeview to Arlington. David averaged 50 mph and Matthew averaged 40 mph. Matthew left one hour earlier than David. They arrived in Arlington at the same time. How long did it take each man to make the trip? __________

6. Ellen hiked uphill at the rate of 3.5 mph. She hiked down the same hill at the rate of 4 mph. The round trip took one hour. How far did she hike? __________

7. Two ships left the same Pacific island at the same time heading in opposite directions. One was traveling 40 mph and the other 30 mph. In how many hours were they 280 miles apart? __________

8. Two buses, 600 miles apart, travel toward each other at rates of 35 mph and 40 mph. How far has the slower bus gone when the buses pass each other? __________

9. A person makes a 400-mile trip, part by car and part by plane. Each part takes two hours. The rate of the plane is three times faster than the rate of the car. How far does the person drive? __________

Using Algebra to Solve Other Word Problems

You can use algebra to solve other types of word problems. For example, in simple interest word problems, you can use algebra to find the amount of money invested at different interest rates. In such problems, you are usually given the rates of interest for each account and the total amount of money invested. You may be asked to find the amount of money invested at each rate.

The equation used to solve interest problems is:

interest = principal × rate × time, or
$I = prt$

Examples

A. An investor had $50,000 to invest. Part of the investment earned 6% interest and part earned 10% interest. Total annual interest earnings were $4,200. How much was invested at each rate?

Step 1 The key words are **had $50,000 to invest, earned 6%, earned 10%**, and **total earnings were $4,200.**

Step 2 Let x = the amount of money invested at the first rate of 6%; $50{,}000 - x$ = the amount of money invested at the second rate of 10%.

To solve the problem, set up a chart. Remember to change the percents to their decimal or fraction equivalent.

Investment	Principal	Rate	Time	Interest
First	x	0.06	1	$0.06x$
Second	$50{,}000 - x$	0.10	1	$0.1(50{,}000 - x)$
Combined	$50,000		1	$4,200

 For more information, see Book 6, pages 128–143.

Step 3 The following equation can be set up based on the information in the chart.

$$\begin{aligned}
\text{interest at 6\%} + \quad \text{interest at 10\%} &= \$4{,}200 \\
0.06x + 0.1(\$50{,}000 - x) &= \$4{,}200 \\
0.06x + \quad \$5{,}000 - 0.1x &= \$4{,}200 \\
-0.04x + \quad \$5{,}000 &= \$4{,}200 \\
-0.04x &= \$4{,}200 - \$5{,}000 \\
-0.04x &= -\$800 \\
x &= \$20{,}000
\end{aligned}$$

$20,000 was invested at 6%.
The amount invested at 10% was $50,000 – $20,000, or 30,000.

Step 4 To check, add the two amounts. You should get $50,000.

$$\$20{,}000 + \$30{,}000 = \$50{,}000 \checkmark$$

Some word problems refer to ages in the past or in the future. To express a past age, subtract from the present age. To express a future age, add to the present age.

MATH HINT

When more than one age is to be represented, let x be the age you know the least about. This will make writing expressions for the other ages in terms of x much easier.

B. Ellen's mother is 3 times as old as Ellen is now. In 10 years, her mother will be twice as old as Ellen is then. How old are Ellen and her mother today?

Step 1 The key words are **3 times as old as, now, in 10 years,** and **twice as old as Ellen is then.**

Step 2 Use a chart to organize your information.

Time	Ellen's Age	Mother's Age
Now In 10 years	x $x + 10$	$3x$ $2x + 10$

Step 3 Set up the equation based on the information in the chart.

mother's age in 10 years = twice Ellen's age in 10 years

$$\begin{aligned}
3x + 10 &= 2(x + 10) \\
3x + 10 &= 2x + 20 \\
3x - 2x &= 20 - 10 \\
x &= 10
\end{aligned}$$

Ellen is 10 years old. Her mother's age is $3x$, or 3(10).
Ellen's mother is 30 years old today.

Step 4 To check, substitute 10 for x in the equation.

$$3x + 10 = 2(x + 10)$$
$$3(10) + 10 = 2(10 + 10)$$
$$30 + 10 = 2(20)$$
$$40 = 40 \checkmark$$

C. A cashier has tallied the change in the register. He has 5 more dimes than nickels. He also has 8 quarters. If the total amount in the register is \$4, what is the value of the dimes and nickels?

Step 1 The key words are **5 more dimes than nickels, 8 quarters,** and **total amount is \$4.**

Step 2 If x = the number of nickels, then
$x + 5$ = the number of dimes

Set up a chart:

Type of Coin	Number of Coins	Value of Coin	Total Value of Coins
nickel	x	0.05	$0.05x$
dime	$x + 5$	0.10	$0.10(x + 5)$
quarter	8	0.25	0.25(8)
		Total Value	4.00

Step 3 The equation is:

$$0.05x + 0.10(x + 5) + 0.25(8) = \$4.00$$
$$0.05x + 0.10x + 0.50 + \$2.00 = \$4.00$$
$$0.15x + \$2.50 = \$4.00$$
$$0.15x = \$4.00 - \$2.50$$
$$0.15x = \$1.50$$
$$x = 10$$

There are 10 nickels in the register. The value of the nickels is 10(\$0.05), or \$0.50.

There are $x + 5$, or 15 dimes. The value of the dimes is 15(\$0.10), or \$1.50.

Step 4 To check, add the values of the coins.

$$0.05(10) + 0.10(15) + 0.25(8) = \$4.00$$
$$\$0.50 + \$1.50 + \$2.00 = \$4.00$$
$$\$4.00 = \$4.00 \checkmark$$

Practice

Solve the following problems.

1. Mary invested some money at 6% interest and an amount $1,000 less than that at 5% interest. If the total interest was $335, how much was invested at each rate? ________________

2. Jane is twice as old as Pat is now. In 15 years, the sum of their ages will be 60. How old are they today? ________________

3. The local bike races charge admittance fees of $4.50 for adults and $2 for children. One day, the receipts were $1,572.50. If twice as many children as adults watched the races, how many adult's and how many child's tickets were sold? ________________

4. Simon invested $10,000 in stocks and bonds. He earned 6% interest on the stocks and 8% interest on the bonds. The total interest earned on the bonds was $40 less than the interest earned on the stocks. How much was invested in each? ________________

5. John is 36 years old and has two daughters. The older daughter is five more than twice the age of the younger daughter. In three years, the sum of the ages of the two daughters will be 26. How old are the daughters? ________________

6. David is 3 times as old as Matthew is now. In 15 years, the sum of their ages will be 70. How old are David and Matthew today? ________________

7. An amount of money was invested at 5%, and $1,500 more than this amount was invested at 7%. If the combined annual interest was $585, how much was invested at each rate? ________________

LIFE SKILL

Work Problems

Work problems involve people or machines working together to complete a job.

To solve work problems, use the formula

part done = rate × time, or
$p = rt$

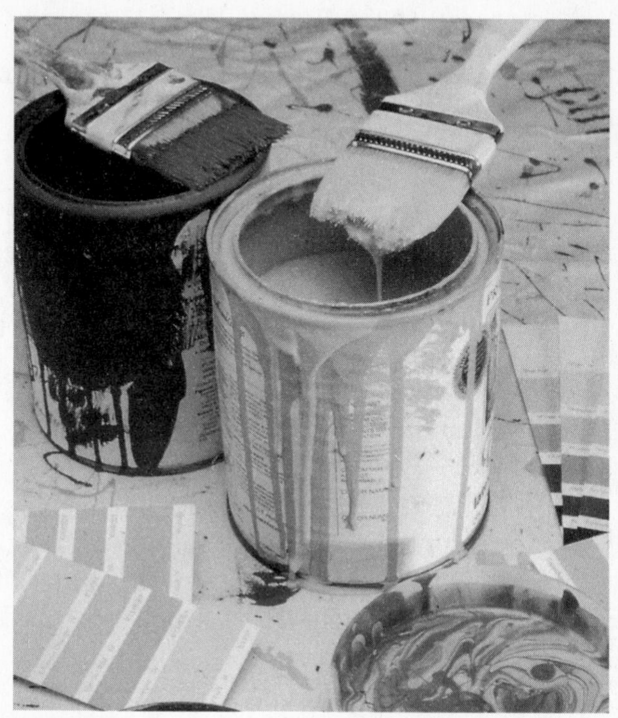

It takes Susan 5 hours to paint a room. It takes Melinda 4 hours to paint the same room. If Susan and Melinda work together, how long will it take them to paint the room?

A table will help you organize the information that you are given.

Let x = the time it takes for Susan and Melinda to paint the room together.

To determine the rates for each person to do the job, express the rate in terms of one unit of time. Then add the rates for each worker to find the combined rate.

Person	Rate (per hour)	Time	Part done
Susan	$\frac{1}{5}$ or 0.2	5 hours	1
Melinda	$\frac{1}{4}$ or 0.25	4 hours	1
Combined	0.45	x	1

Taking the information from the table, solve the problem using the work equation.

$$
\begin{aligned}
p &= rt \\
1 &= 0.45x \\
\frac{1}{0.45} &= \frac{0.45x}{0.45} \\
2.22 &= x
\end{aligned}
$$

It will take Susan and Melinda approximately 2.2 hours to paint the room together.

Solve the following problems.

1. Ellen can sew a dress in 3 hours. Connie can sew the same dress in 5 hours. How long will it take to finish the dress if they work together?

____

2. Printer A takes 5 hours to print a report. If Printer A and Printer B work together, a report can be completed in 2 hours 44 minutes. How long does it take Printer B to print the report working alone?

____

3. A tank can be filled in 10 hours using one pipe. It takes a smaller pipe 15 hours to fill the tank. How long will it take to fill the tank if both pipes are used?

____

4. JoEllen can paper a small room in 5 hours. It takes Susan 8 hours to paper the same size room. How long will it take them to paper the room together?

____

UNIT

11

Posttest

Solve the following problems.

1. When you triple a number and subtract 14, you get two-thirds of the original number. What is the original number?

   ____

2. The sum of three consecutive odd integers is 249. Find the numbers.

   ____

3. A car travels 2 hours at a rate of 40 mph. How long will it take to travel the same distance at a rate of 60 mph?

   ____

4. A basketball team scored 69 points in 41 shots. Only two-point shots and one-point foul shots were made. How many of each type of shot did the team make?

   ____

5. Stuart is 8 years older than his brother Carl. Twenty years ago, Carl's age was one-third of Stuart's age then. How old are Stuart and Carl today?

   ____

6. An amount of money was invested at 10%, and $1,500 less than this amount was invested at 8%. If the combined annual interest was $690, how much money was invested at each rate?

   ____

UNIT

12

Exponents, Multiplication, and Factoring

Pretest

Simplify.

1. $\frac{2^4}{2^2}$ ________

2. $p^4 \cdot 2p^8$ ________

3. $x^2(3x^2)^4$ ________

4. $\frac{t^{10}}{t^4}$ ________

5. $2x(y^3 + 8y^4)$ ________

6. $5t^2(4t + 2t^3 + 5t^3)$ ________

7. $(x + 5)(x + 1)$ ________

8. $(s - 12)(s + 12)$ ________

9. $(x^2 + 4)(3x - 7)$ ________

10. $(14 - x)(1 + x)$ ________

11. $(y - 13)(y + 13)$ ________

12. $(k - 21)(k + 22)$ ________

Factor each of the following.

13. $x^2 + 9x + 14$

____

14. $x^2 + 14x + 48$

____

15. $x^2 - 8x + 16$

____

16. $b^2 + 8b + 7$

____

17. $4x^2 + 48x + 144$

18. $x^2 - 25$

Solve each of the following equations.

19. $x^2 - 16 = 0$

20. $x^2 + 20x + 100 = 0$

21. $x^2 - 5x = -6$

22. $(x + 3)(2x - 1) = 9$

Problem Solving

Solve the following problems.

23. A softball league has 12 teams. What is the total number of games to be played? Use the formula $n^2 - n = N$ where n is the number of teams and N is the number of games played.

24. The sides of a square are increased by 3 inches on each side. The area of the new square is 81 square inches. Find the length of a side of the original square.

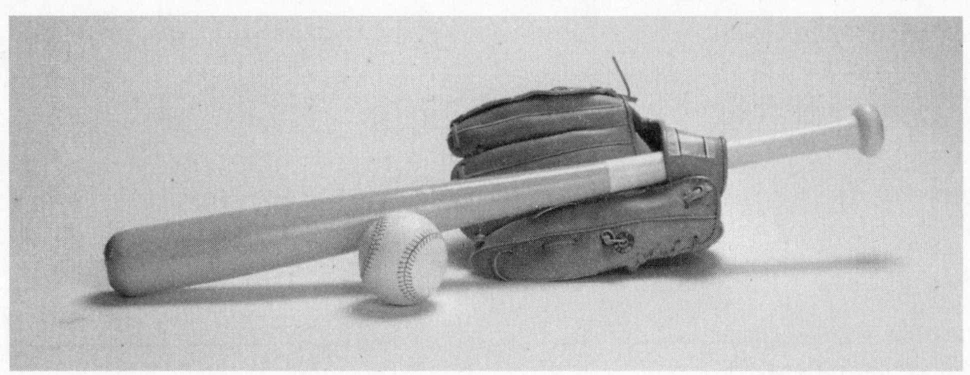

Exponents

In the expression $3^4 = 81$, the number 3 is the **base** and the number 4 is the **exponent**. The exponent tells us that the base, 3, is used 4 times as a factor.

Examples

A. The base, 3, multiplies itself 4 times.
The base, 2, multiplies itself 5 times.
The base, $\frac{1}{5}$, multiplies itself 4 times.

$3^4 = 3 \cdot 3 \cdot 3 \cdot 3 = 81$
$2^5 = 2 \cdot 2 \cdot 2 \cdot 2 \cdot 2 = 32$
$\left(\frac{1}{5}\right)^4 = \frac{1}{5} \cdot \frac{1}{5} \cdot \frac{1}{5} \cdot \frac{1}{5} = \frac{1}{625}$

B. When a variable is used as a factor, the multiplication looks like this.

$x^4 = x \cdot x \cdot x \cdot x$
$y^2 = y \cdot y$

C. To multiply two quantities with the same base, **add** the exponents.

$2^3 \cdot 2^6$
$2 \cdot 2 \cdot 2 \quad 2 \cdot 2 \cdot 2 \cdot 2 \cdot 2 \cdot 2$
2^{3+6}
$2^9 = 512$

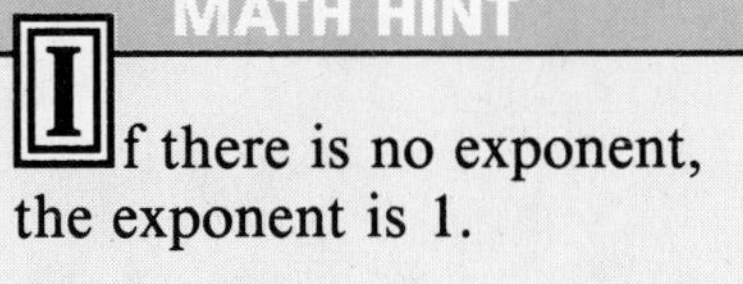
MATH HINT

If there is no exponent, the exponent is 1.

$x^4 \cdot x$
$x \cdot x \cdot x \cdot x \quad x$
x^{4+1}
x^5

D. To divide two quantities with the same base, **subtract** the exponents.

$\frac{2^5}{2^3} = 2^{5-3} = 2^2 = 4$

MATH HINT

Notice how the factors cancel.
$\frac{\not{2} \cdot \not{2} \cdot \not{2} \cdot 2 \cdot 2}{\not{2} \cdot \not{2} \cdot \not{2}} = \frac{2^2}{1} = 2^2 = 4$

For more information, see Book 6, pages 146–148; 155.

E. When a quantity raised to a power is again raised to a power, keep the base and multiply the exponents.

$(3^2)^3 = 3^2 \cdot 3^2 \cdot 3^2 = 3^{2+2+2} = 3^6$

or

$(3^2)^3 = 3^{2 \cdot 3} = 3^6 = 729$

$(x^2y^2)^4 = x^2y^2 \cdot x^2y^2 \cdot x^2y^2 \cdot x^2y^2 = x^8y^8$

or

$(x^2y^2)^4 = x^{2+2+2+2} \cdot y^{2+2+2+2} = x^8y^8$

or

$(x^2y^2)^4 = x^{2 \cdot 4} \cdot y^{2 \cdot 4} = x^8y^8$

Practice

Find the value of each of the following.

1. $3 \cdot 3 \cdot 3$ __________

2. $2 \cdot 2 \cdot 2 \cdot 2$ __________

3. 7^2 __________

4. 4^3 __________

5. $\left(\frac{1}{2}\right)^2$ __________

6. $(0.03)^3$ __________

Write each of the following with exponents.

7. $x \cdot x \cdot x \cdot x \cdot x$ __________

8. $y \cdot y \cdot y$ __________

9. $2 \cdot 2 \cdot 2 \cdot 2 \cdot 2 \cdot 2$ __________

10. $\left(\frac{1}{3}\right)\left(\frac{1}{3}\right)\left(\frac{1}{3}\right)$ __________

Simplify.

11. $a^3 \cdot a^3$ __________

12. $b^4 \cdot b^6$ __________

13. $2^2 \cdot 2^3$ __________

14. $b \cdot b^4$ __________

15. $x^6 \cdot x^{15}$ __________

16. $y^9 \cdot y^3$ __________

17. $\frac{a^3}{a^2}$ __________

18. $\frac{y^6}{y^4}$ __________

19. $\frac{x^6}{x^8}$ __________

20. $\frac{y^4}{y^7}$ __________

21. $\frac{a^4}{a}$ __________

22. $\frac{x^7}{x^3}$ __________

23. $\frac{y}{y^5}$ __________

24. $\frac{a}{a}$ __________

25. $\frac{x^{16}}{x^{15}}$ __________

26. $\frac{c^6}{c^{13}}$ __________

27. $(x^3)^2$ __________

28. $(y^6)^3$ __________

29. $(y^8)^4$ __________

30. $(3^2)^2$ __________

31. $(x^4)^7$ __________

32. $(a^4)^4$ __________

33. $(x^2y^2)^3$ __________

34. $(a^3b^4)^2$ __________

35. $\frac{3^4}{3^2}$ __________

36. $x^5(x^7)$ __________

37. $(2x^6)^2(3x^2)^5$ __________

38. $\frac{(x^6)^3}{x^4}$ __________

39. $3xy^2 + (2xy)^2$ __________

40. $2(xy)^2(x^2y^3)^4$ __________

Problem Solving

Solve the following problems.

41. A perfect square is a number that can be found by multiplying another number by itself. For example, 9 is a perfect square because it is the square of 3. Other perfect squares are 1, 4, 16, and 25. Find all of the perfect squares between 100 and 200.

____

42. A perfect cube is a number that can be found by multiplying another number by itself three times. For example, 8 is a perfect cube because $2 \times 2 \times 2 = 8$.

$$2^3 = 8$$

Other perfect cubes are 1, 27, 64, and 125. Find the perfect cubes between 200 and 500.

____

LIFE SKILL

Scientific Notation

One way of writing very large numbers is to use scientific notation. A number is in scientific notation if it is written as a number between one and ten multiplied by a power of ten.

Write 4,400 in scientific notation.

4,400 in scientific notation is 4.4×10^3.

Using scientific notation and the rules for multiplying and dividing exponents, you can multiply very large numbers or very small numbers easily.

Solve $(7.5 \times 10^5) \times (9.8 \times 10^4)$.

$(7.5 \times 10^5) \times (9.8 \times 10^4) = (7.5 \times 9.8) \times (10^5 \times 10^4)$

$= 73.5 \times 10^9$ Rewrite in scientific notation.

$= 7.35 \times 10^{10}$

Solve 2.4×10^9 divided by 2×10^3.

$$\frac{2.4 \times 10^9}{2 \times 10^3} = \frac{2.4 \times 10^{9-3}}{2}$$

$$= 1.2 \times 10^6$$

Write each of the following in scientific notation.

1. 5,000

____

2. 34,876

____

3. 758

____

4. 5,800

____

Solve.

5. $(4.5 \times 10^3)(2.3 \times 10^2)$

6. $\frac{(1.2 \times 10^5)}{(3.6 \times 10^5)}$

7. $\frac{(9 \times 10^4)}{(4.5 \times 10^3)}$

8. $(3.2 \times 10^4)(8 \times 10^2)$

Multiplication and the Distributive Property

When multiplying terms, first multiply the numbers together; then multiply the variables. When the terms have parentheses, multiply everything inside the parentheses by the number directly in front of them.

Examples

A. $(5x^3)(2x^2) = (5 \cdot 2) \cdot x^3 \cdot x^2$
$= 10x^{3+2}$
$= 10x^5$

MATH HINT

Remember to add exponents to multiply variables.

B. $2(y + 3) = (2 \cdot y) + (2 \cdot 3)$
$= 2y + 6$

C. $-3(x^2 + 4) = -3(x^2) + [(-3)(4)]$
$= -3x^2 + (-12)$
$= -3x^2 - 12$

D. $y^2(y^2 - 5) = (y^2 \cdot y^2) - (y^2)(5)$
$= y^4 - 5y^2$

E. $7x^2(x^3 - 5x^2) = 7x^2(x^3) - (7x^2)(5x^2)$
$= 7x^5 - 35x^4$

Practice

Simplify.

1. $(3x^2)(5x^3)$

____

2. $4(3x + 2)$

____

3. $(-2y^3)(3y^4)$

____

4. $-3(x^2 + 9)$

____

5. $2x(4x^2 + 2)$

____

6. $-5x^3(6x + 5)$

____

 For more information, see Book 6, pages 150–151.

7. $7a(a - 4)$ ____________

8. $16c^2(2c^3 - 3c)$ ____________

9. $-9y^3(3y + 2)$ ____________

10. $6(x^2 + 2x + 1)$ ____________

11. $-9(3x^2 + 4x + 3)$ ____________

12. $3y^2(x^2 + x + 1)$ ____________

13. $(-y)(-y)(-y)$ ____________

14. $-2b^3(4b^2 - 3bc)$ ____________

15. $5ax(6a^2 + 2ax + 4)$ ____________

16. $-9x(3y^3 + 2y^2 + 3)$ ____________

Multiplying Two-Term Expressions Using F.O.I.L.

To multiply a two-term expression by a two-term expression, multiply each term of the first by each term of the second. You can also use the **F.O.I.L.** method of multiplication.

Examples

A. Simplify $(3x + 4)(x + 5)$.

$$\begin{array}{rl} 3x + 4 & \\ \underline{x + 5} & \\ 15x + 20 & \text{multiply 5 times } (3x + 4) \\ \underline{3x^2 + 4x} & \text{multiply } x \text{ times } (3x + 4) \\ 3x^2 + 19x + 20 & \text{add the products} \end{array}$$

B. Simplify $(x + 3)(x + 6)$.

F. O. I. L.

$(x + 3)(x + 6) = x^2 + 6x + 3x + 18$
$= x^2 + 9x + 18$

First	$x \cdot x = x^2$
Outside	$x \cdot 6 = 6x$
Inside	$3 \cdot x = 3x$
Last	$3 \cdot 6 = 18$

C. Simplify $(n - 2)(n + 5)$.

$(n - 2)(n + 5) = n^2 + 5n - 2n - 10$
$= n^2 + 3n - 10$

First	$n \cdot n = n^2$
Outside	$n \cdot 5 = 5n$
Inside	$-2 \cdot n = -2n$
Last	$-2 \cdot 5 = -10$

D. Simplify $(2x + 8)(3x - 5)$.

$(2x + 8)(3x - 5) = 6x^2 - 10x + 24x - 40$
$= 6x^2 + 14x - 40$

First	$2x \cdot 3x = 6x^2$
Outside	$2x \cdot (-5) = -10x$
Inside	$8 \cdot 3x = 24x$
Last	$8 \cdot (-5) = -40$

 For more information, see Book 6, pages 152–154.

Practice

Multiply.

1. $(x + 3)(x - 2)$ __________

2. $(x + 4)(x + 6)$ __________

3. $(y - 5)(y + 5)$ __________

4. $(m - 10)(m + 2)$ __________

5. $(ab + 3)(ab - 5)$ __________

6. $(xy - 8)(xy - 8)$ __________

7. $(4 + x)(10 + x)$ __________

8. $(7 - t)(7 + t)$ __________

9. $(2x + 5)(3x - 2)$ __________

10. $(3m - 12)(2m + 4)$ __________

11. $(2x^2 - 3)(2x^2 + 9)$ __________

12. $(3x - 6)^2$ __________

13. $(-1 + 3p)(2 + 5p)$ __________

14. $2x(x - 3)(x + 2)$ __________

15. $-(x - 4y)(x + y)$ __________

16. $(3m - 2n)(m + n)$ __________

The Greatest Common Factor

A **factor** is a whole number that will divide into a group of numbers with no remainder. In multiplication, the numbers being multiplied are factors.

MATH HINT

The number 1 is a factor of all numbers.

Compare the factors of 24 and 36.

Factors of 24		**Factors of 36**	
	$1 \cdot 24$		$1 \cdot 36$
	$2 \cdot 12$		$2 \cdot 18$
	$3 \cdot 8$		$3 \cdot 12$
	$4 \cdot 6$		$4 \cdot 9$
			$6 \cdot 6$

Circle the factors that are common to both 24 and 36.

24 1, 2, 3, 4, 5, 6, 12, 24
36 1, 2, 3, 4, 9, 12, 18, 36

The numbers 1, 2, 3, 4, and 12 are common to both lists of factors. The largest factor they have in common is 12. This number is called the **greatest common factor (GCF)**.

Variable expressions can have a GCF.

$x^5 = x \cdot x \cdot x \cdot x \cdot x$
$x^7 = x \cdot x \cdot x \cdot x \cdot x \cdot x \cdot x$
GCF $= x^5$

$c^4 = c \cdot c \cdot c \cdot c$
$c^2 = c \cdot c$
GCF $= c^2$

MATH HINT

When variable expressions have the same base, the GCF is the base with the smallest exponent.

 For more information, see Book 6, pages 160–161.

Examples

To factor the GCF in an algebraic expression, follow these steps:

A. Factor $4x + 6$.

Step 1	Find the GCF of the terms.	The GCF of $4x$ and 6 is 2.
Step 2	Divide each term by the GCF to find the contents of the parentheses.	$\frac{4x}{2} = 2x \quad \frac{6}{2} = 3$
Step 3	Show the factors.	$2(2x + 3)$

B. Factor $36y^5 - 12y^2$.

Step 1 The GCF is $12y^2$.

Step 2 $\frac{36y^5}{12y^2} = 3y^3 \quad -\frac{12y^2}{12y^2} = -1$

Step 3 $12y^2(3y^3 - 1)$

MATH HINT

Subtract exponents when dividing by the GCF.

C. Factor $3x^2 + 24xy + x^3$.

Step 1 The GCF is x.

Step 2 $\frac{3x^2}{x} = 3x \quad \frac{24xy}{x} = 24y \quad \frac{x^3}{x} = x^2$

Step 3 $x(3x + 24y + x^2)$

Practice

Find the GCF for each set.

1. 8, 12 ________

2. 18, 24 ________

3. 12, 16 ________

4. 18, 24, 30 ________

5. 7, $14x$ ________

6. $5x$, $10y$ ________

7. $6xy$, $8xy$ ________

8. $6x^2$, $4x$ ________

9. $10x^2$, $15x^3$ ________

10. $8x^2y^2, -16x^2y$

11. $50x^3y, 75x^5y^3$

12. $3(2x - y), 5(2x - y)$

Factor each of the following.

13. $8x + 12$

14. $12y - 18$

15. $3x + 3y$

16. $4y - 12$

17. $9x + 12xy$

18. $10x^2 - 15x$

19. $12x^3 + 16x^2$

20. $15x^3 - 15xy$

21. $25a^2b^2 - 100a^3b^2$

22. $2(a + b) + 3a(a + b)$

Factoring

Remember the pattern that F.O.I.L. created.

$$(x + 4)(x + 6) = \overset{F}{x^2} + \overset{O}{6x} + \overset{I}{4x} + \overset{L}{24}$$

$$= \overset{F}{x^2} + \overset{OI}{(6 + 4)x} + \overset{L}{24}$$

$$= x^2 + 10x + 24$$

To factor a three-term expression, follow these steps that show the reverse of the F.O.I.L. process.

Factor $x^2 + 10x - 24$.

Step 1 Find the numbers whose product is the last term and whose sum is the coefficient of the middle term.

Factors of 24	Sum of Factors
24, 1	$24 + 1 = 25$
−24, −1	$-24 + -1 = -25$
8, 3	$8 + 3 = 11$
−8, −3	$-8 + -3 = -11$
12, 2	$12 + 2 = 14$
−12, −2	$-12 + -2 = -14$
(6, 4)	($6 + 4 = 10$)
−6, −4	$-6 + -4 = -10$

Step 2 Place the factors of the first term into the parentheses. $(x \quad)(x \quad)$

Step 3 Fill in the two factors found in Step 1. The factors can be listed in any order. $(x + 6)(x + 4)$

Step 4 Check by multiplying with F.O.I.L. $(x + 6)(x + 4) = x^2 + 10x + 24$

Examples

A. Factor $x^2 + 7x + 10$.

Step 1

Factors of 10	Sum of Factors
10, 1	$10 + 1 = 11$
−10, −1	$-10 + -1 = -11$
(5, 2)	($5 + 2 = 7$)
−5, −2	$-5 + -2 = -7$

Step 2 $(x \quad)(x \quad)$

For more information, see Book 6, pages 160–166.

Step 3 $(x + 5)(x + 2)$

Step 4 $(x + 5)(x + 2) = x^2 + 7x + 10$

B. Factor $x^2 - 11x + 18$.

Step 1

Factors of 10	**Sum of Factors**
18, 1	$18 + 1 = 19$
−18, −1	$-18 + -1 = -19$
9, 2	$9 + 2 = 11$
(−9, −2)	$(-9 + -2 = -11)$
6, 3	$6 + 3 = 9$
−6, −3	$-6 + -3 = -9$

Step 2 $(x \quad)(x \quad)$

Step 3 $(x - 9)(x - 2)$

Step 4 $(x - 9)(x - 2) = x^2 - 11x + 18$

MATH HINT

You cannot factor all expressions. An expression may not have a GCF or the sum of the factors of the third term may not equal the second term.

Practice

Fill in the chart with the factors that will give each sum and product. The first one has been done for you.

	Sum	Product	Factors
1.	−7	12	−3, −4
2.	−10	21	
3.	2	−35	
4.	5	−24	
5.	12	11	

Factor.

6. $x^2 - 6x + 5$ ____________

7. $x^2 + 3x + 2$ ____________

8. $x^2 - 8x + 7$ ____________

9. $x^2 - 2x - 15$ ____________

10. $x^2 + 12x + 20$

11. $x^2 - x - 30$

12. $x^2 - 10x + 21$

13. $x^2 - 5x - 14$

14. $x^2 + 14x + 49$

15. $x^2 - 8x + 16$

16. $x^2 + 2x - 99$

17. $x^2 - 8x + 12$

Notice the pattern that develops when we multiply the following numbers. These become two-term expressions called the **difference of two squares**.

$$(x - 5)(x + 5) = x^2 + 5x - 5x - 25 = x^2 - 25$$
$$(x + 8)(x - 8) = x^2 - 8x + 8x - 64 = x^2 - 64$$

When we reverse the order of F.O.I.L., these two-term expressions can be factored as the sum and difference of their square roots.

Examples

C. $x^2 - 49$
$(x \quad)(x \quad)$
$(x - 7)(x + 7)$

D. $x^2 - 100$
$(x \quad)(x \quad)$
$(x - 10)(x + 10)$

Practice

Factor.

18. $x^2 - 4$

19. $x^2 - 9$

20. $x^2 - 36$

21. $x^2 - 144$

22. $x^2 - \frac{1}{9}$

23. $x^2 - 0.04$

24. Factor: $x^4 - 81$. (Hint: You may have to factor more than once.)

25. Factor: $2x^4 - 18$. (Hint: Make sure to take the GCF out first.)

26. Factor $3x^2 + x - 10$. (Hint: Try several groups inside the parentheses until you find the right match of factors.)

27. Factor $4x^2 + 10x + 6$. (Hint: Make sure to take the GCF out first.)

L E S S O N 7 3

Solving Equations and Factoring

To solve equations that have three-term expressions, follow these steps.

Solve $x^2 + 4x - 5 = 0$.

Step 1 Factor the expression.

$(x + 5)(x - 1) = 0$

Step 2 Set each factor equal to zero.

$x + 5 = 0$ $\quad$ $x - 1 = 0$

Step 3 Solve for each value.

$$\begin{array}{rr} x + 5 = & 0 \\ -5 & -5 \\ \hline x = & -5 \end{array} \qquad \begin{array}{rr} x - 1 = & 0 \\ +1 & +1 \\ \hline x = & 1 \end{array}$$

Step 4 Check each solution.

$$\begin{array}{l} x^2 + 4x - 5 = 0 \\ (-5)^2 + 4(-5) - 5 = 0 \\ 25 - 20 - 5 = 0 \\ 25 - 25 = 0 \\ 0 = 0 \end{array} \qquad \begin{array}{l} x^2 + 4x - 5 = 0 \\ (1)^2 + 4(1) - 5 = 0 \\ 1 + 4 - 5 = 0 \\ 5 - 5 = 0 \\ 0 = 0 \end{array}$$

Examples

Find the solution for *x*.

A. $x^2 - 3x + 2 = 0$

$(x - 2)(x - 1) = 0$

$$\begin{array}{rr} x - 2 = & 0 \\ +2 & +2 \\ \hline x = & 2 \end{array} \qquad \begin{array}{rr} x - 1 = & 0 \\ +1 & +1 \\ \hline x = & 1 \end{array}$$

Check:

$$\begin{array}{l} x^2 - 3x + 2 = 0 \\ (2)^2 - 3(2) + 2 = 0 \\ 4 - 6 + 2 = 0 \\ 6 - 6 = 0 \\ 0 = 0 \end{array} \qquad \begin{array}{l} x^2 - 3x + 2 = 0 \\ 1 - 3(1) + 2 = 0 \\ 1 - 3 + 2 = 0 \\ 3 - 3 = 0 \\ 0 = 0 \end{array}$$

B. $6x^2 - 24x = 0$

$6x(x - 4) = 0$

$$\frac{6x}{6} = \frac{0}{6} \qquad \begin{array}{rr} x - 4 = & 0 \\ +4 & +4 \\ \hline x = & 4 \end{array}$$

$x = 0$

Check:

$$\begin{array}{l} 6x^2 - 24x = 0 \\ 6(0) - 24(0) = 0 \\ 0 - 0 = 0 \\ 0 = 0 \end{array} \qquad \begin{array}{l} 6x^2 - 24x = 0 \\ 6(4)^2 - 24(4) = 0 \\ 6(16) - 24(4) = 0 \\ 96 - 96 = 0 \\ 0 = 0 \end{array}$$

For more information, see Book 6, pages 167–168.

Practice

Solve each equation. Check your solution.

1. $x^2 + 8x + 15 = 0$

2. $x^2 + 7x + 12 = 0$

3. $x^2 - 6x + 9 = 0$

4. $x^2 - 5x = 0$

5. $y^2 - 4y - 45 = 0$

6. $5x^2 - 6x = 0$

7. $x^2 + 10x + 25 = 0$

8. $x^2 - 16 = 0$

9. $x^2 - 7x = 0$

10. $x^2 + 9x + 8 = 0$

Problem Solving

Solve the following problems.

11. Eight more than the square of a number is six times the number. Find the number. (Hint: Move everything except for zero to one side of the equal sign before factoring.)

12. The area of a square is 5 more than the perimeter. Find the length of the side. (Hint: Move everything except for zero to one side of the equal sign before factoring.)

Problem Solving—Using Factoring

The steps you have learned to help solve word problems can be used with word problems that deal with factoring. Follow these steps:

Step 1 Read the problem and underline the key words. These words will usually relate to some mathematical reasoning.

Step 2 Make a plan to solve the problem. Ask yourself, Should I add, subtract, multiply, divide, round, or compare? You may have to do more than one of these operations for the same problem.

Step 3 Find the solution. Use your math knowledge to find your answer.

Step 4 Check your answer. Ask yourself, Is the answer reasonable? Did you find what you were asked for?

When solving problems with factoring, first identify what is given and what is needed.

Examples

A. The product of two positive consecutive even integers is 168. What are the integers?

Step 1 Read the problem and identify the key words.
The key words are **product** and **positive consecutive even integers**.

Step 2 Make a plan to solve the problem. Set up the variables.

Let x = first even integer.
Let $x + 2$ = next even integer.
What do you know? The product of the integers is 168.
Set up the equation: $x(x + 2) = 168$

For more information, see Book 6, pages 169–170.

Step 3 Find the solution.

$$x(x + 2) = 168$$
$$x^2 + 2x - 168 = 0$$
$$(x + 14)(x - 12) = 0$$

$$x + 14 = 0 \text{ or } x - 12 = 0$$
$$x = -14 \text{ or } x = 12$$

Because the first integer must be positive, we can eliminate -14. The two positive consecutive even integers are 12 and 14.

Step 4 Check your answer.
$12(14) = 168$
$168 = 168$

MATH HINT

It is always important to check your answers when factoring. Sometimes one or both of the answers do not make sense in the problem even though mathematically they are correct.

B. A gardener uses 72 feet of wire fencing to enclose a rectangular plot for a garden. The area of the garden is 260 square feet. Find the length and width of the garden.

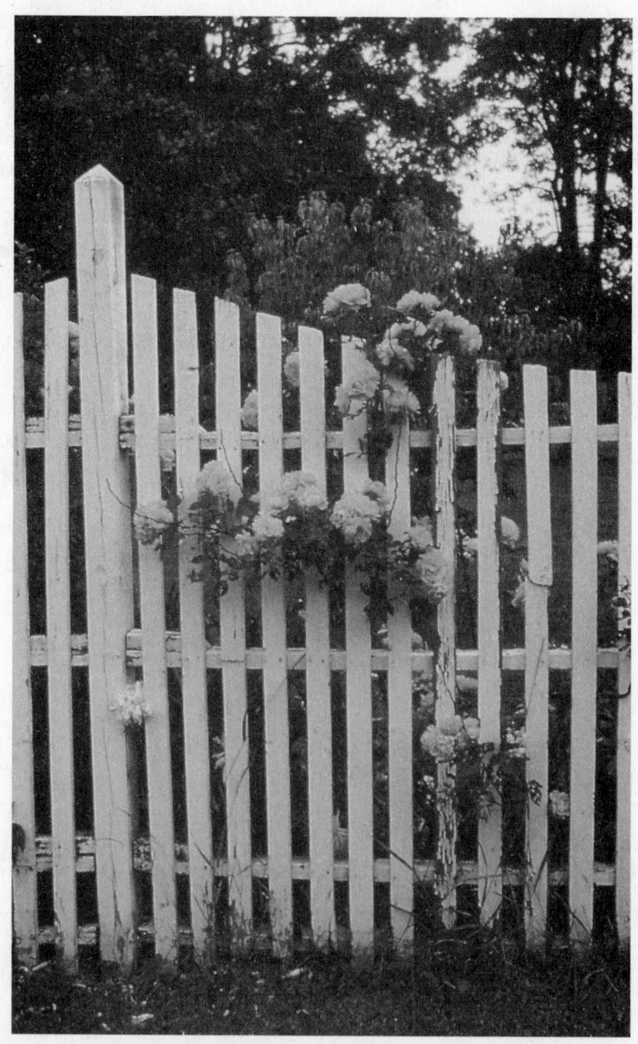

Step 1 Read the problem and identify the key words. The key words are **enclose**, **area**, **length**, and **width**.

Step 2 Make a plan to solve the problem. You want to find the length and width of the plot. The perimeter is 72 feet and the area of the garden is 260 square feet.
To find the dimensions of the plot, use these equations:

$P = 2l + 2w$,
where P is the perimeter, l is the length, and w is the width.

$A = lw$,
where A is the area, l is the length, and w is the width.

Step 3 Find the solution of one of the equations.

$$A = lw$$
$$260 = lw$$
$$\frac{260}{w} = l$$

Substitute $\frac{260}{w}$ for l in the other equation.

$$P = 2l + 2w$$
$$72 = 2\left(\frac{260}{w}\right) + 2w$$
$$72 = \frac{520}{w} + 2w$$
$$72w = 520 + 2w^2$$
$$0 = 2w^2 - 72w + 520$$
$$0 = 2(w^2 - 36w + 260)$$
$$0 = 2(w - 26)(w - 10)$$

Remove the fraction by multiplying both sides of the equation by w.

$$\begin{array}{rr} w - 26 = & 0 \\ +26 & +26 \\ \hline w = & 26 \end{array} \qquad \begin{array}{rr} w - 10 = & 0 \\ +10 & +10 \\ \hline w = & 10 \end{array}$$

Because length is usually longer than width, the length must be 26 feet and the width 10 feet.

Step 4 Check your answer by substituting your answers into the equations.

$$A = lw$$
$$260 = 26(10)$$
$$260 = 260 \checkmark$$

$$P = 2l + 2w$$
$$72 = 2(26) + 2(10)$$
$$72 = 52 + 20$$
$$72 = 72 \checkmark$$

Problem Solving

Solve the following problems.

1. The product of two consecutive odd integers is 143. Find the integers.

____

2. The sum of the squares of two consecutive odd integers is 74. Find the integers.

____

3. The base of a triangle is 10 feet greater than the height. The area of the triangle is 28 square feet. Find the height and the base. (Hint: $A = \frac{1}{2}bh$)

____

4. The sum of the squares of two consecutive odd negative numbers is 130. Find the numbers.

____

UNIT 12

Posttest

Simplify.

1. $y^2(4x^2y)$

2. $d^4 \cdot d^8 \cdot d \cdot d^6$

3. $\frac{42^{24}}{42^{21}}$

4. $\left(\frac{z^8}{z^6}\right)^2$

5. $x(x^3 + 2x^4)$

6. $2v^2(v + 2v^3 + v^3)$

7. $(y + 9)(y - 1)$

8. $(h - 7)(h + 7)$

9. $(4x^2 + 2)(5x - 3)$

10. $(20 - x)(1 - x)$

11. $(y - 21)(y + 21)$

12. $(c - 2z)(c + 12z)$

Factor each of the following.

13. $x^2 + 7x - 18$

14. $x^2 + 4x - 21$

15. $x^2 - 8x + 16$

16. $m^2 + 6m - 7$

17. $5n^2 + 45n + 70$

18. $x^2 - 121$

Solve each of the following.

19. $16 = x^2$

20. $x^2 - 9x + 14 = 0$

21. $6x^2 - 4x = 10$

22. $(5x + 4)(x - 1) = 2$

Problem Solving

Solve the following problem.

23. At a meeting, there are 40 people. How many handshakes are possible? Use the formula $N = \frac{(n^2 - n)}{2}$, where N is the number of handshakes and n is the number of people present.

UNIT

13

Graphing and Inequalities

Pretest

For problems 1–4, locate and label each point on the grid.

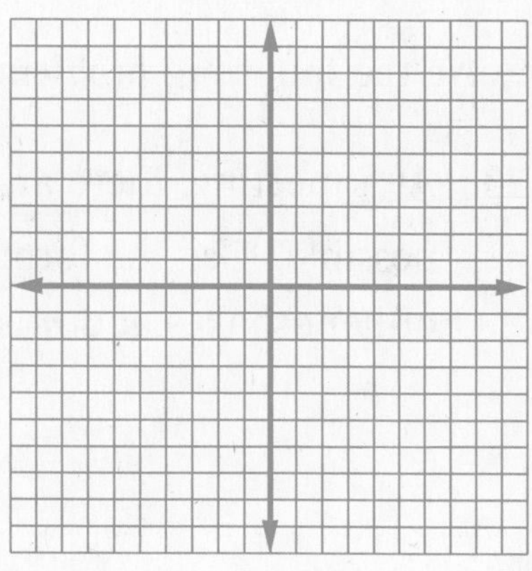

1. Point $A = (4, 1)$

2. Point $B = (-3, 2)$

3. Point $C = (5, 0)$

4. Point $D = (8, 2)$

Find the distance between the pairs of points.

5. (4, 3) and (4, 0) ______

6. (0, 5) and (4, 2) ______

Find the slopes of the lines through the given points.

7. (2, 6), (1, 5) ______

8. (2, 0), (4, 4) ______

Find the value of y for each ordered pair.

9. $2x + y = 8$

(1, ______) (4, ______) (2, ______)

10. $y = 2x + 1$

(2, ______) (0, ______) (5, ______)

11. Graph $2x + y = 8$ and $y = 2x + 1$ on the grid.

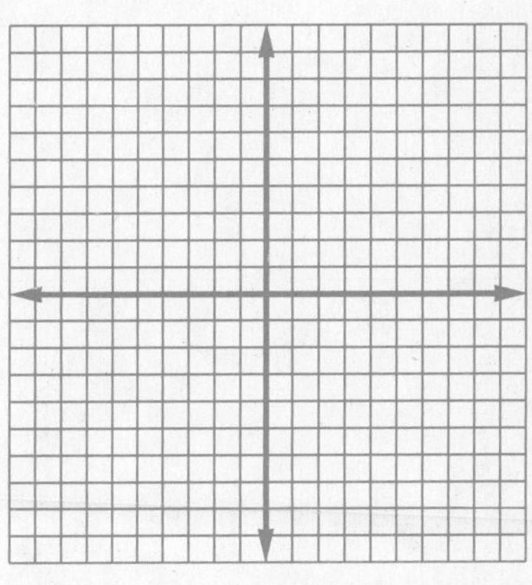

Graph each of the solution sets on a number line.

12. $x \leq -4$

13. $x > 0$

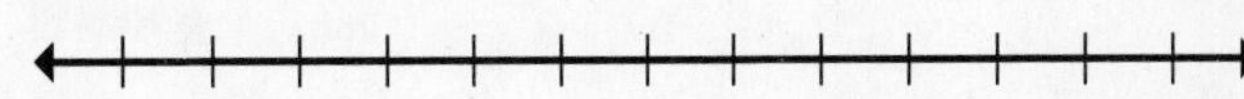

14. x is between -2 and 3

15. $x > -4$

Solve each inequality and graph the solution.

16. $x - 2 > 8 + 2x$

17. $5x < 4 - 10 + 7x$

18. $-9 + 2x \leq x - 30$

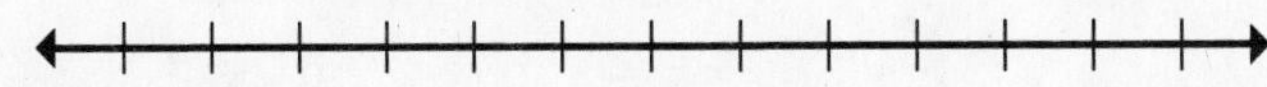

19. $8x < 3x - 6 + 26$

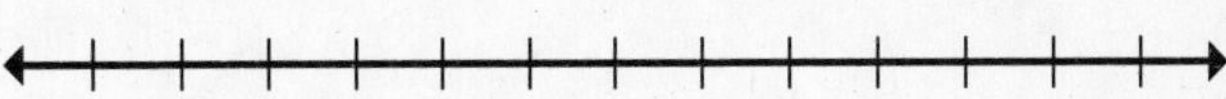

Problem Solving

Write the inequality for each problem and solve. Then check your solution.

20. The sum of a number and 15 is less than four times the number. Find all the numbers that make this statement true.

______

21. Simone reports that it costs at least $3 for each call to a customer. It costs $0.75 plus $0.45 per minute for the calls. What is the average time that Simone spends on the phone with a customer?

______

Points on a Grid, Distance, and Slope

A **plane**, or a flat surface, can be divided into four parts by a horizontal number line, called the **x-axis** and a vertical number line called the **y-axis**. The two lines intersect at a point called the **origin**.

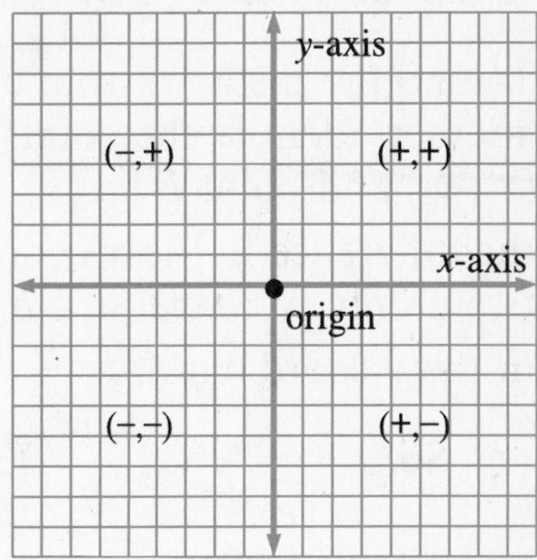

The location of any point in the plane can be identified by a pair of numbers called the **coordinates** (or **ordered pair**) of the point. In the coordinate (x, y), the first number tells the horizontal distance of the point from the origin. If this number is positive, the point is to the right of the origin. If it is negative, the point is to the left of the origin. The second number tells the vertical distance from the origin. If the second number is positive, the point is above the origin; if it is negative, the point is below the origin.

Examples

A. Locate and label the point $(-4, 2)$ on the grid. Place your pencil on the origin $(0, 0)$, and count left 4 spaces. Then count up 2 spaces.

B. Locate and label the point $(5, -4)$ on the grid. Place your pencil on the origin, and count right 5 spaces. Then count down 4 spaces.

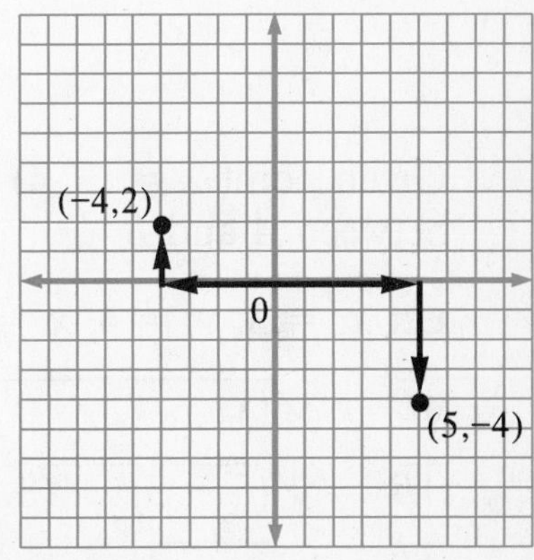

MATH HINT

To locate the point, always begin by moving along the x-axis.

For more information, see Book 6, pages 175–180.

To find the distance between two points on a graph, you can use the Pythagorean theorem. This theorem says that in a right triangle, the sum of the squares of the shorter sides (the legs) is equal to the square of the longer side (the hypotenuse).

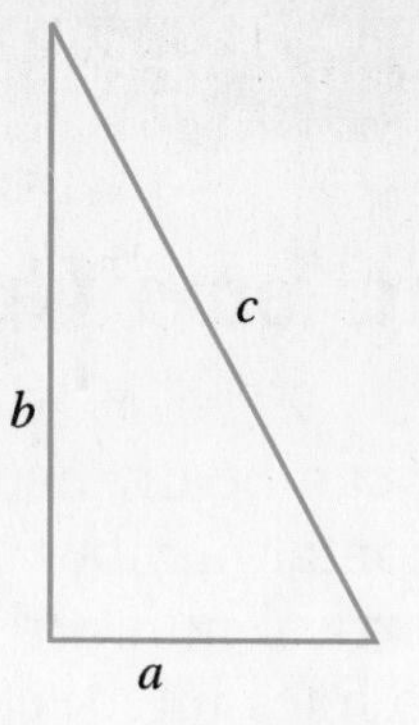

$c^2 = a^2 + b^2$

or $c = \sqrt{a^2 + b^2}$

To find the distance from one point to another, identify the points. Let point A be (x_1, y_1) and point B be (x_2, y_2). The side a is the difference in the x coordinates of the two points, $x_2 - x_1$. Side b is the difference in the y coordinates, $y_2 - y_1$.

So, $\overline{AB} = \sqrt{(x_2 - x_1)^2 + (y_2 - y_1)^2}$

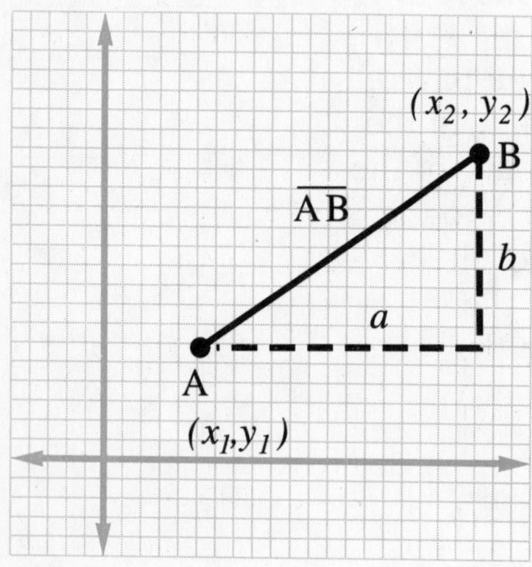

Examples

C. Given point $A = (3, 4)$ and point $B = (5, 8)$, find the distance between A and B.

Let $x_1 = 3$, $y_1 = 4$, $x_2 = 5$, and $y_2 = 8$

$\overline{AB} = \sqrt{(x_2 - x_1) + (y_2 - y_1)^2}$

$\overline{AB} = \sqrt{(5 - 3)^2 + (8 - 4)^2}$ Simplify inside the parentheses first.

$\overline{AB} = \sqrt{2^2 + 4^2}$

$\overline{AB} = \sqrt{4 + 16}$

$\overline{AB} = \sqrt{20}$

$\overline{AB} =$ about 4.47

The distance between point A and point B is about 4.47 units.

D. Given point $A = (-2, 6)$ and point $B = (7, 4)$, find the distance between A and B.

Let $x_1 = -2$, $y_1 = 6$, $x_2 = 7$, and $y_2 = 4$

$$\overline{AB} = \sqrt{(x_2 - x_1)^2 + (y_2 - y_1)^2}$$
$$\overline{AB} = \sqrt{[7 - (-2)]^2 + (4 - 6)^2}$$
$$\overline{AB} = \sqrt{9^2 + (-2)^2}$$
$$\overline{AB} = \sqrt{85}$$
$$\overline{AB} = \text{about } 9.22$$

The distance between point A and point B is about 9.22 units.

Slope is the steepness or slant of a hill. A steep hill has a large slope. On a coordinate graph, a line is said to have a slope. The slope of a line is the change in the y coordinates over the change in the x coordinates. The slope of the line can be positive, negative, zero, or undefined.

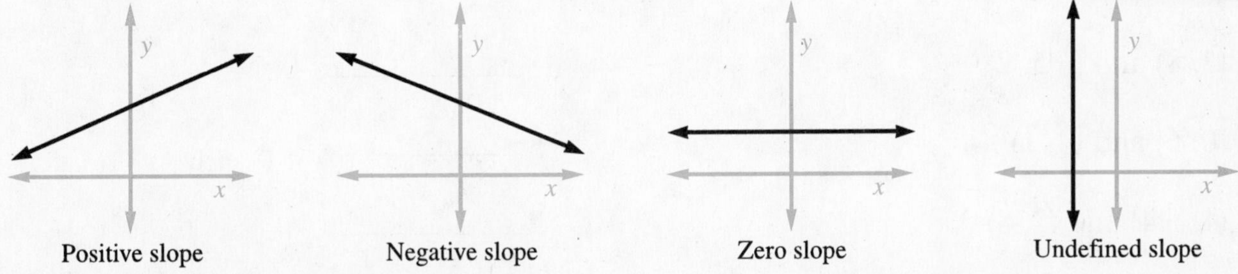

If a line passes through the points $A(x_1, y_1)$ and $B(x_2, y_2)$, the slope is found by using the formula

$$slope = \frac{y_2 - y_1}{x_2 - x_1}, \text{ or } \frac{\text{change in } y}{\text{change in } x}$$

Example

E. Find the slope of a line passing through the points $A(2, 4)$ and $B(6, 3)$.

$x_1 = 2$, $y_1 = 4$ and $x_2 = 6$, $y_2 = 3$

$$slope = \frac{y_2 - y_1}{x_2 - x_1}$$
$$slope = \frac{3 - 4}{6 - 2}$$
$$slope = \frac{-1}{4} = -\frac{1}{4}$$

The slope is $-\frac{1}{4}$, which means the line goes down to the right.

Practice

Locate and label each of the following points on the grid.

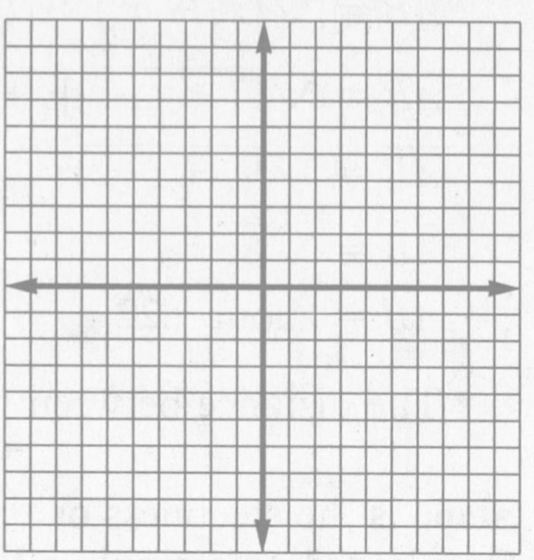

1. Point $A = (1, 2)$
2. Point $B = (3, 8)$
3. Point $C = (-5, 2)$
4. Point $D = (5, 0)$
5. Point $E = (1, -4)$

Use the formula $\sqrt{(x_2 - x_1)^2 + (y_2 - y_1)^2}$ to find the distance between each pair of points.

6. (1, 2) and (3, 8) __________
7. (3, 8) and (−5, 2) __________
8. (0, 6) and (4, 1) __________
9. (1, −4) and (2, 8) __________

Find the slopes of the lines through the given points.

10. (2, 3), (6, 5) __________
11. (0, 0), (2, 3) __________
12. (3, 4), (1, 5) __________
13. (2, −1), (2, 5) __________
14. (−3, −5), (−6, −5) __________
15. (0, −6), (0, 5) __________

Solutions to and Graphs of Equations

A **solution** to an equation is the numerical value that makes the equation true.

$$\begin{array}{r} x + 4 = 9 \\ -4 = -4 \\ \hline x = 5 \end{array}$$

Until now, all of your equations have contained only one variable. Let's look at an equation with two variables.

$x + y = 5$

If $x = 2$, then $y = 3$ because $2 + 3 = 5$.
If $x = 3$, then $y = 2$ because $3 + 2 = 5$.
If $x = 0$, then $y = 5$ because $0 + 5 = 5$.

If you change the value of x, the value of y changes. The number of solutions is unending. One way of expressing the solutions to this equation is by ordered pairs. The solution for $x + y = 5$ includes the following ordered pairs.

(2, 3) (3, 2) (0, 5)

Examples

A. Find the ordered pairs that make $y - 2x = 0$ a true equation.

(1, ______) If $x = 1$, then $y - 2(1) = 0$. So, $y = 2$.

(0, ______) If $x = 0$, then $y - 2(0) = 0$. So, $y = 0$.

(2, ______) If $x = 2$, then $y - 2(2) = 0$. So, $y = 4$.

B. Find the ordered pairs that make $y = x + 1$ a true equation.

(1, ______) If $x = 1$, then $y = 1 + 1$. So, $y = 2$.

(0, ______) If $x = 0$, then $y = 0 + 1$. So, $y = 1$.

(−1, ______) If $x = -1$, then $y = -1 + 1$. So, $y = 0$.

For more information, see Book 6, pages 184–195.

To graph the equation of a line, you need to know three points on the line. These points are solutions to the equation.

> **MATH HINT**
>
> A table of values helps to organize your ordered pairs.

Examples

C. Graph $x + y = 5$.

Find three ordered pairs.

x	y
2	3
0	5
−2	7

(2, 3) (0, 5) (−2, 7)

Locate the three points on the grid.

Draw a straight line through the three points. Any point on the line is a solution to the equation. The line is called a graph of the equation $x + y = 5$.

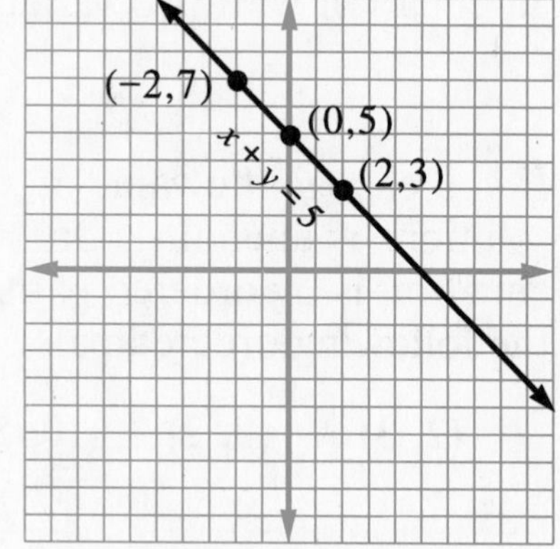

Sometimes an equation is not known. A graph can be drawn using a table of known values. These graphs only predict solutions because the graph may not be a straight line.

D. An owner of a florist shop finds that the price of his flowers determines how many he sells. He records the information by using ordered pairs where the first value is the number of dozens of flowers he sells and the second value is the price per dozen.

The table below shows his record.

x	y
50	\$8
100	\$4
125	\$2

Plot the points and predict how many dozen flowers will be purchased when he prices the flowers at \$6 a dozen.

Looking at the graph, you can predict that 75 dozen flowers will be purchased at \$6 a dozen.

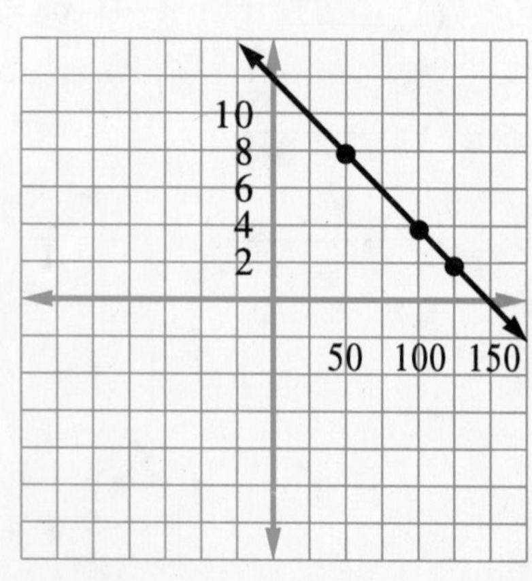

Graphing two equations on the same coordinate graph is another way to compare data. Graph the first equation. Then graph the second equation.

MATH HINT

You may want to put the graphs in different colors to help you keep them separate.

E. Graph $x + y = 5$ and $2x + y = 8$ on the same graph.

$x + y = 5$	
x	y
0	5
5	0
2	3

$2x + y = 8$	
x	y
1	6
2	4
4	0

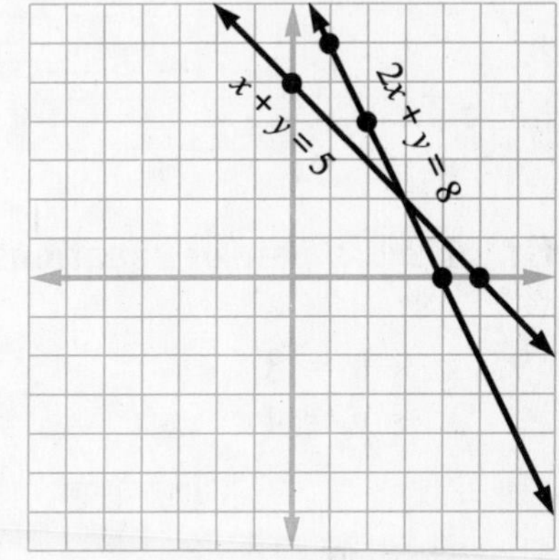

The point of intersection (3, 2) is a solution of both equations.

Practice

Find the value of y for each of the ordered pairs.

1. $3x + y = 8$

(1, ___) (0, ___) (2, ___)

2. $y = 2x - 2$

(−1, ___) (0, ___) (1, ___)

3. $x + 3y = 19$

(19, ___) (16, ___) (25, ___)

4. $x + y = 5$

(7, ___) (2, ___) (0, ___)

5. $y = 4x + 1$

(2, ___) (5, ___) (3, ___)

6. $y = \frac{2(x)}{3} - 1$

(3, ___) (9, ___) (0, ___)

7. Graph the equation given in exercise 1.

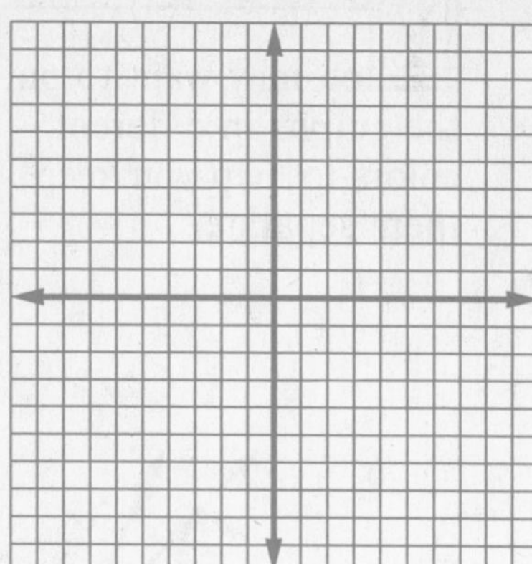

8. Graph the equation given in exercise 4.

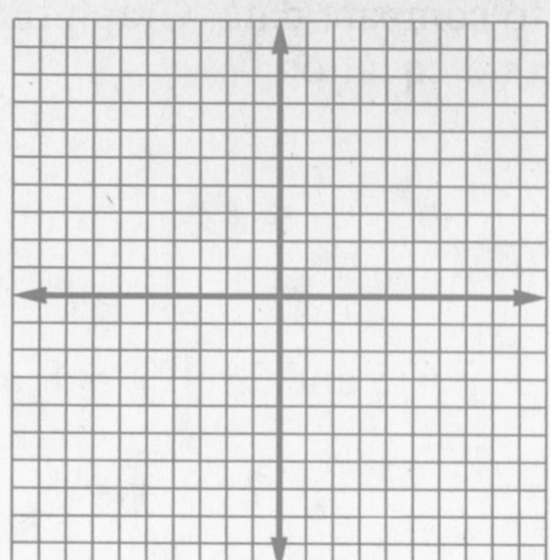

Graph each pair of equations on the same grid.

9. $x + y = 8$
   $x - y = 4$

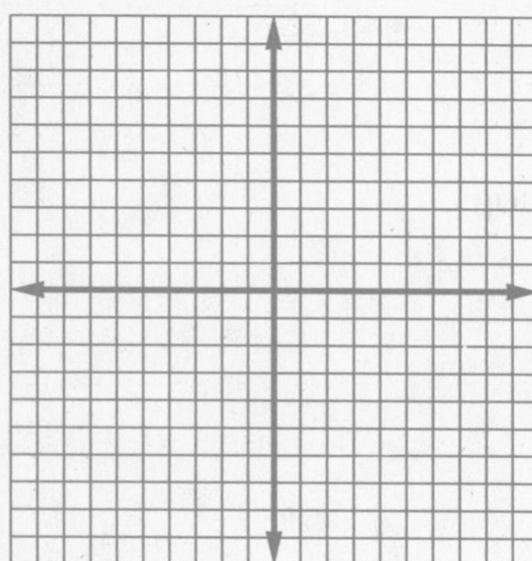

10. $y = 3x - 5$
    $y = 3x - 1$

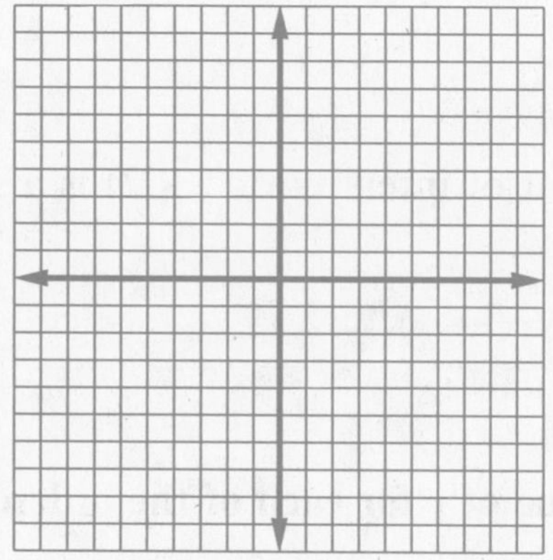

11. $y = \frac{1}{2}x + 3$
    $y = 4x - 5$

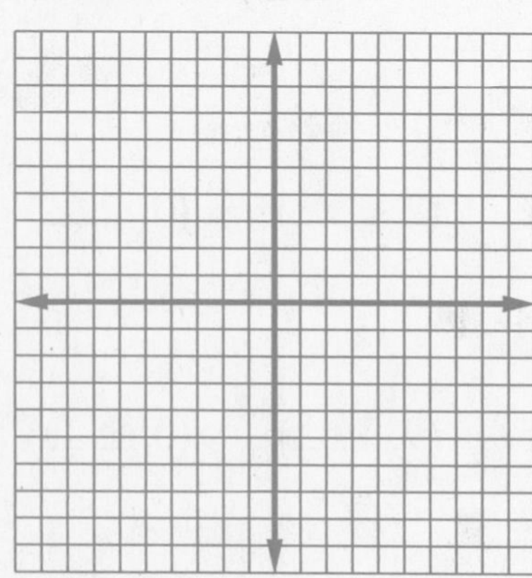

12. $2x + 3y = 9$
    $4x + 6y = 18$

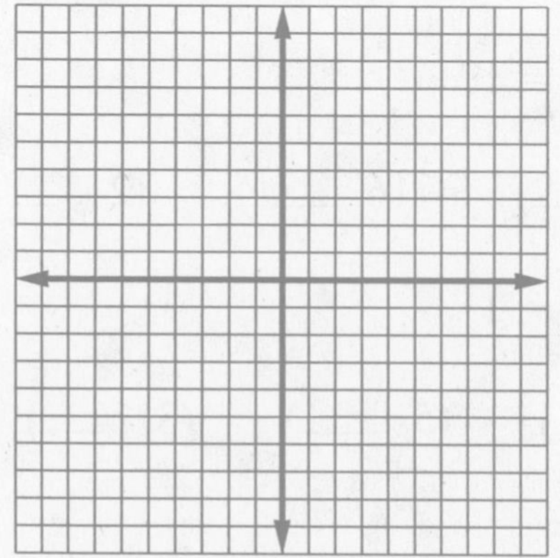

LESSON 77

Inequalities

An expression written with $>$, $\geq$, $<$, or $\leq$ is called an **inequality**.

- $>$ means "is greater than," so $5 > 3$ is read: Five is greater than 3.
- $\geq$ means "is greater than or equal to," so $x \geq 3$ is read: x is greater than or equal to 3.
- $<$ means "is less than," so $1 < 3$ is read: One is less than 3.
- $\leq$ means "less than or equal to," so $x \leq 3$ is read: x is less than or equal to 3.

You can use a number line to show the solution of an inequality. If the solution includes the endpoint, use a closed circle. If the solution does not include the endpoint, use an open circle.

Examples

A. Graph the solution of $x \geq -1$.
The endpoint -1 is included in the answer, so place a **closed** circle on the number -1. Next shade everything to the right of -1.

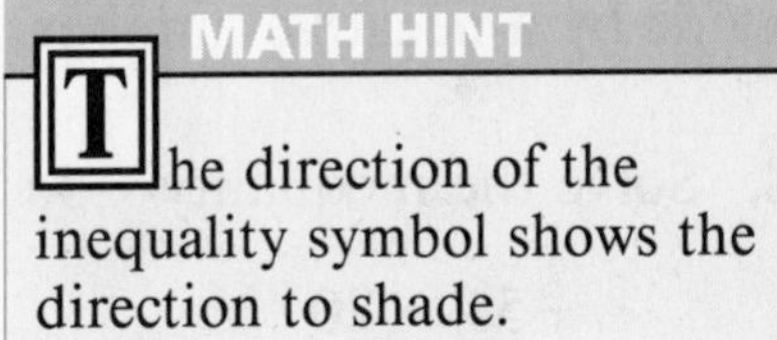

MATH HINT
The direction of the inequality symbol shows the direction to shade.

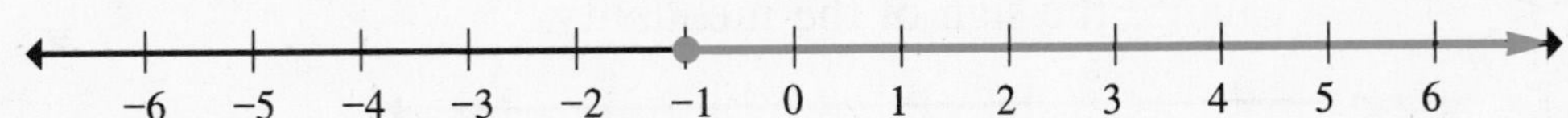

B. Graph the solution of $x < 4$.
The endpoint 4 is not included in the answer, so place an **open** circle on the number 4. Next shade everything to the left of 4.

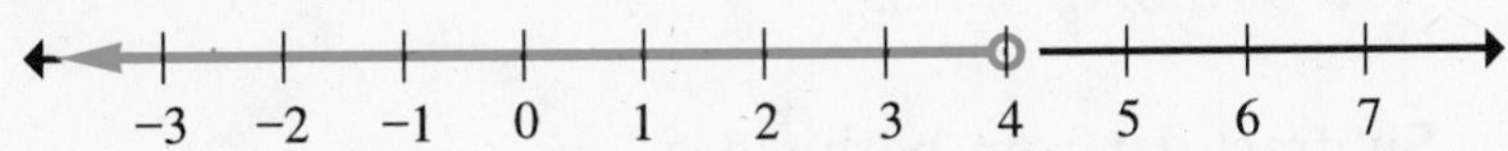

Solving an inequality is much the same as solving an equation. You are finding values for the variables that will make the statement true. Inequalities can have more than one solution.

For more information, see Book 6, pages 201–207.

Examples

C. Solve the inequality $x + 4 < 14$. Graph the solution.

$$\begin{array}{rcl} x + 4 < & 14 & \text{Subtract 4 from both sides.} \\ -\ 4 & -\ 4 & \\ \hline x < & 10 & \end{array}$$

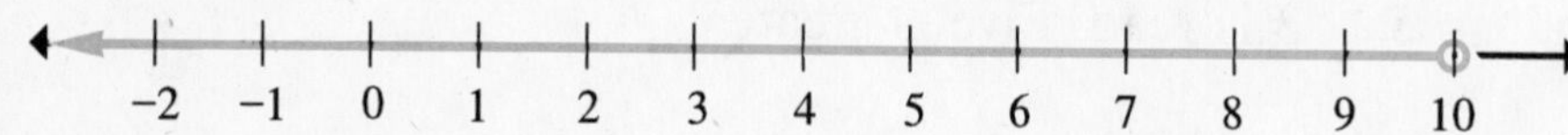

D. Solve the inequality $4p - 3.2 > 3p + 0.7$. Graph the solution.

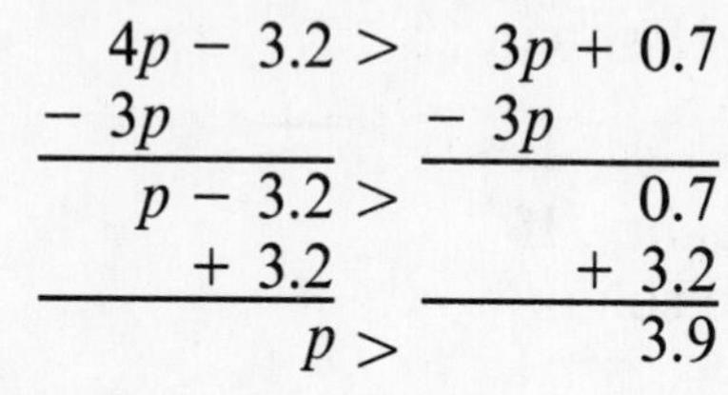

$$\begin{array}{rcl} 4p - 3.2 > & 3p + 0.7 & \\ -\ 3p & -\ 3p & \text{Subtract } 3p \text{ from both sides.} \\ \hline p - 3.2 > & 0.7 & \text{Combine like terms.} \\ +\ 3.2 & +\ 3.2 & \text{Add 3.2 to both sides.} \\ \hline p > & 3.9 & \end{array}$$

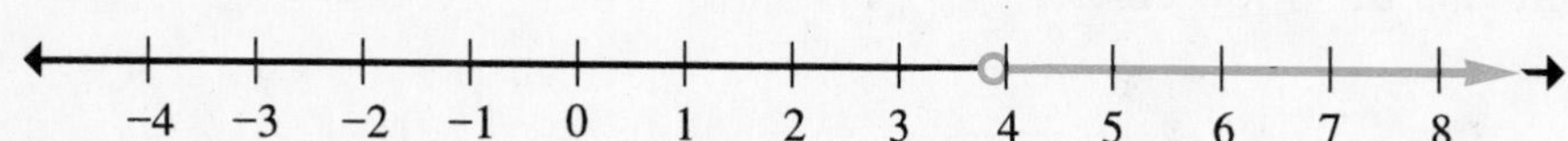

Solving an inequality is different from solving equations only when you multiply or divide by a negative number. When you multiply or divide by a negative number, the direction of the inequality changes.

E. Solve the inequality $-5c > 20$. Graph the solution.

$$-5c > 20$$

$$\frac{-5c}{-5} > \frac{20}{-5} \qquad \text{Divide both sides by } -5.$$

$$c < -4 \qquad \text{Change the sign of the inequality.}$$

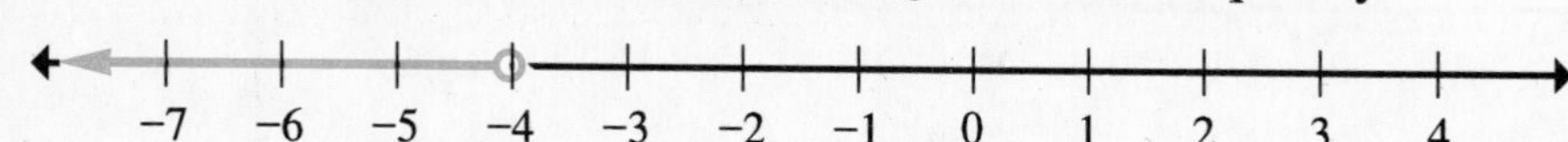

F. Solve the inequality $2x - 8 > 4x$. Graph the solution.

$$\begin{array}{rcl} 2x - 8 > & 4x & \\ +\ 8 & +\ 8 & \text{Add 8 to both sides.} \\ \hline 2x > & 4x + 8 & \\ -\ 4x > & -\ 4x & \text{Subtract } 4x \text{ from both sides.} \\ \hline -\ 2x > & 8 & \end{array}$$

$$\frac{-2x}{-2} > \frac{8}{-2} \qquad \text{Divide both by } -2.$$

$$x < -4 \qquad \text{Change the sign of the inequality.}$$

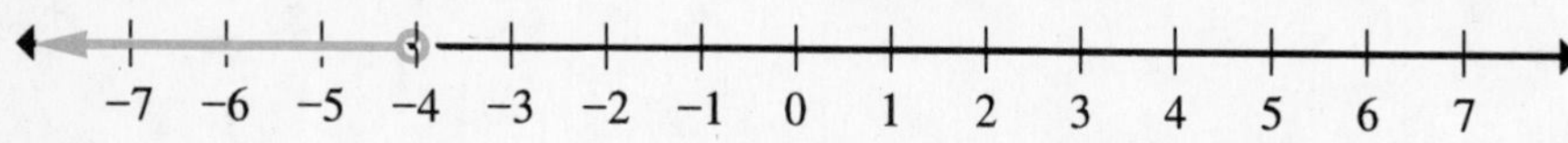

Practice

Solve each of the following. Graph the solution.

1. $x \leq 2$

2. $x > -2$

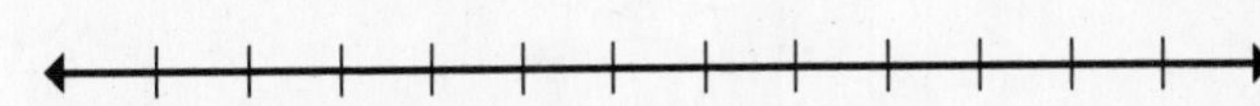

3. x is between 2 and 5

4. $x < -3$

5. $x > 4$

6. x is between -1 and 3

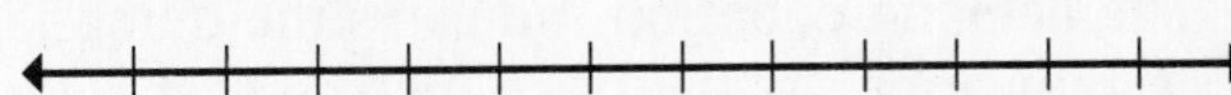

Solve each inequality. Graph the solution.

7. $9x - 21 > 80 + 8x$

8. $6x > 14 - 1 + 7x$

9. $-19 + x < 2x - 35$

10. $2x + 18 < 3x - 6$

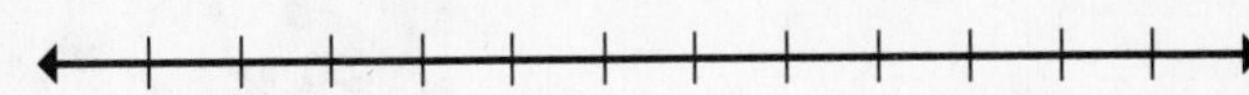

11. $4x - 8 < 3x$

12. $5x - 9 < 4 + 4x$

13. $3x + 9 > x + 7$

14. $-5x > 25$

LIFE SKILL

Supply and Demand Equations

In business, you can use two equations to help determine the price of an item. These equations are (1) the consumer's demand equation and (2) the seller's supply equation.

The **demand** equation indicates the demand of an item at a certain price. As the price increases, the consumer wants less of the item.

The **supply** equation indicates the amount of items the seller is willing to supply at a given price. As the price increases, the seller is willing to supply more items.

The best situation for both the consumer and the seller is that the demand and the supply are equal. To see where this point is located, graph the two equations on the same grid, and note the meeting point. This is the price and the amount of the item needed. This is called the **equilibrium point**.

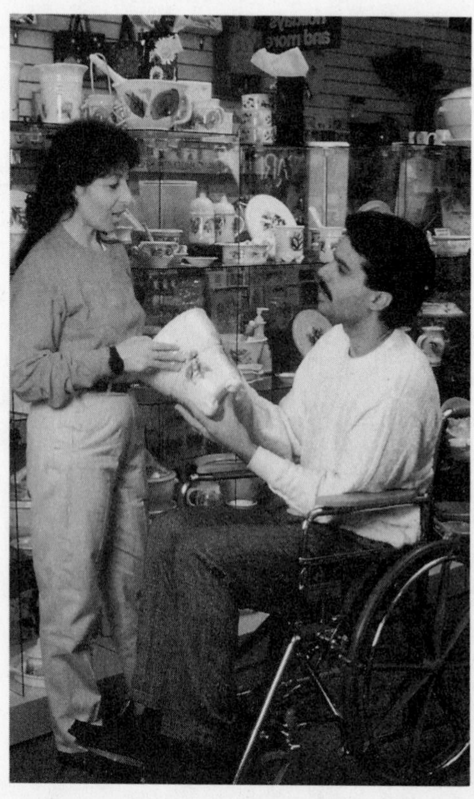

Let's look at an example. In the example, the number of items is in thousands of units.

The demand equation is $-3x + 40 = y$, where x is the price and y is the demand. The supply equation is $4x + 12 = y$, where x is the price and y is the supply.

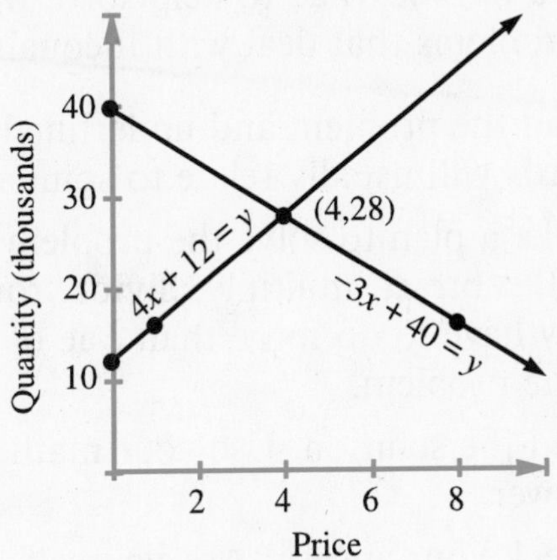

As you can note from the graph, the lines meet at the point (4, 28). This means the price should be \$4 and the seller should supply 28,000 units of the item.

1. If the seller supplies more than 28,000 items, what do you think will happen?

____

2. If the seller supplies less than 28,000 items, what do you think will happen?

____

Problem Solving—Using Inequalities

The steps you have learned to help solve word problems can be used with word problems that deal with inequalities. Follow these steps:

Step 1 Read the problem and underline the key words. These words will usually relate to some mathematical reasoning.

Step 2 Make a plan to solve the problem. Ask yourself, Should I add, subtract, multiply, divide, round, or compare? You may have to do more than one of these operations for the same problem.

Step 3 Find the solution. Use your math knowledge to find your answer.

Step 4 Check your answer. Ask yourself, Is the answer reasonable? Did you find what you were asked for?

The fourth step in solving word problems is to check the solution. Sometimes errors are introduced as we solve problems. It is always important to check your solutions to make sure they are reasonable.

Example

A number divided by 2 is at least -10. Write the inequality and solve. Then check the solution.

Step 1 Read the problem and underline the key words. The key words are **divided by** and **at least**.

Step 2 Make a plan.

Let x = the number.
The inequality can be written $\frac{x}{2} \geq -10$.

Step 3 Find the solution.

$\frac{x}{2} \geq -10$ Multiply both sides by 2.
$\frac{2x}{2} \geq -10(2)$
$x \geq -20$

Step 4 Check the solution.

Is $\frac{-20}{2} \geq -10$?
$-10 \geq -10$ ✓

 For more information, see Book 6, page 208.

Practice

Write the inequality for each problem and solve. Then check your solution.

1. Four times a number is no more than 200. Find all numbers that make this statement true.

____

2. Judy's score is at least 20 points greater than the 53-point record. What is the lowest score that Judy can make?

____

3. A 6-centimeter valve can have a diameter no more than 0.5 greater than or 0.5 less than the standard. What measurements are allowed for the valve?

____

4. A 14-inch valve can have a diameter no more than 0.75 greater than or 0.75 less than the standard. What measurements are allowed for the valve?

____

5. A person earns $300 per week. The deductions taken out each week can be no greater than 20% of his pay. What is the most he could have taken out of his check?

____

6. A canoe trip takes no less than 2 hours 30 minutes and no more than 5 hours. If a camper leaves the canoe area at 1:00 p.m., between what two times will the trip be completed?

____

7. Matthew works for $7.50 an hour. If he is guaranteed at least 35 hours each week, what is his minimum wage for one week?

____

8. Cindy needs to buy a car. She wants to put at most 10 percent of the price as a down payment. The car she likes costs $3,500. What is the most she wants to put down?

____

9. Richard and Carra go out for dinner. They spend $20 for dinner. If they tip between 10 and 20 percent, between what two amounts is the tip?

____

UNIT

13

Posttest

For problems 1–4, locate and label each point on the grid.

1. Point $A = (1, 5)$

2. Point $B = (6, 0)$

3. Point $C = (-1, 2)$

4. Point $D = (-4, -2)$

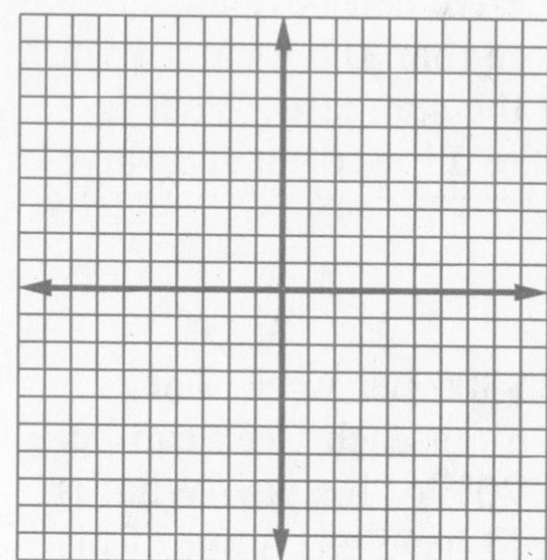

Find the distance between the pairs of points.

5. $(-4, 3)$ and $(2, -1)$ ________

6. $(-2, 5)$ and $(2, 2)$ ________

Find the slopes of the lines through the given points.

7. $(8, 1), (-1, 5)$ ________

8. $(5, 0), (5, 4)$ ________

Find the value of *y* for each ordered pair.

9. $x + 4y = 5$

   (1, ____)   (4, ____)   (2, ____)

10. $y = -2x + 3$

   (2, ____)   (0, ____)   (7, ____)

11. Graph $x + 4y = 5$ and $y = -2x + 3$ on the grid.

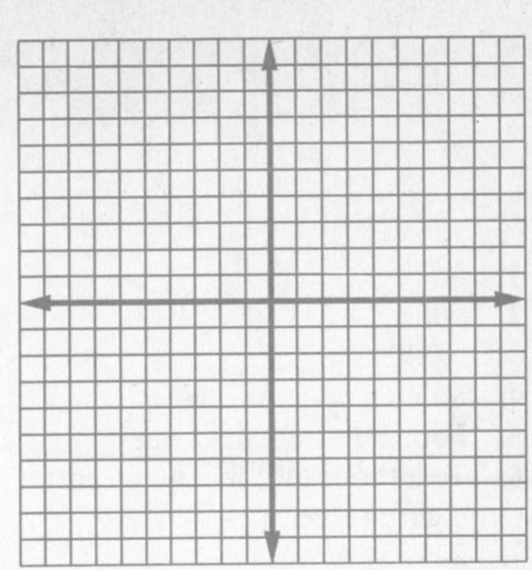

Graph each of the solution sets on a number line.

12. $x > 10$

13. x is between -2 and 6

14. x is at least -2

15. $-5 < x$

Solve each inequality and graph the solution.

16. $5y > -2$

17. $4 + 3x < 28$

18. $6 - 4y > 4 - 3y$

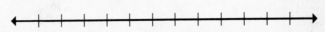

19. $6 - 13x < 4 - 12x$

Problem Solving

Write the inequality for each problem and solve. Then check your solution.

20. One student's quiz scores are 73, 75, 89, and 91. What does the student need on the fifth quiz to get an average of at least 85?

____

21. The area of a square can be no more than 64 square yards. What are the maximum measures for the sides of the square?

____

ANSWERS

Unit 1 Pretest/pages 1-3

1. 80
2. 200
3. 5,000
4. 80
5. Four hundred sixty-one
6. Two thousand, five hundred eighty-three
7. 248
8. 9,014
9. 790
10. 16,000
11. 1,559
12. 2,547
13. 54,702
14. 90,732
15. 505
16. 118
17. 446
18. 282
19. 27,328
20. 12,679
21. 1,566
22. 2,775
23. 5,922
24. 486
25. 98
26. 80
27. 17
28. 25
29. 35
30. 31
31. 20 pounds of cheddar
32. 5,160 cans
33. $142
34. 20 teams

Lesson 1/pages 4-5

1. 100,000
2. 30
3. Nine hundred eighty-two million, one hundred seventy thousand, five hundred thirty-four
4. 982,170,534
5. 80
6. 800
7. 7,000
8. 70
9. 2,000
10. 2
11. Two hundred ninety-four
12. Seventy-one thousand, nine hundred eighty-three
13. Nine hundred fifty-two thousand, seven hundred eighty
14. Thirty-two thousand, sixty
15. Two million, four hundred seventy-five thousand, one
16. 358

Lesson 2/pages 6-7

1. 860
2. 8,100
3. 56,000
4. 105,990
5. 2,200,000
6. 98,000
7. 60,400,000
8. 57,000,000
9. $5,000
10. 52,500

Life Skill/pages 8-9

1. About $5.10
2. About $9.70

Lesson 3/pages 10-12

1. 15
2. 15
3. 49
4. 137
5. 1,361
6. 16,236
7. 3,824
8. 9,145
9. 15,537
10. 90,732
11. $70
12. $410
13. 600 miles
14. 50 miles
15. 32 hours

Lesson 4/pages 13-15

1. 11
2. 14
3. 23
4. 79
5. 24
6. 42
7. 2
8. 16
9. 682
10. 298
11. 181
12. 464
13. 219
14. 13,239
15. 80,773
16. 10 leaves
17. $6,728
18. 37 dozen
19. $30
20. $148

Lesson 5/pages 16-18

1. 156
2. 520
3. 376
4. 6,460
5. 1,971
6. 1,764
7. 689
8. 250
9. 736
10. 0
11. 0
12. 0
13. 50
14. 600
15. 80,000
16. 1,252,092
17. 4,589,786
18. 2,357,432
19. 9,240 baseballs
20. 30 people
21. 363 miles
22. 2,450 calories

Lesson 6/pages 19-20

1. 500
2. Undefined
3. 50
4. 100
5. 5
6. 70
7. 400
8. 200
9. 8
10. 37
11. 17
12. 504
13. 78
14. 38
15. 8,009 R5

Lesson 7/pages 21-22

1. 27
2. 36
3. 625
4. 27
5. 10
6. 32
7. 1
8. 512

9. 15
10. 9
11. 1
12. 25

Lesson 8/pages 23-24

1. 18
2. 1
3. 116
4. 45
5. 28
6. 26
7. 25
8. 18
9. 21
10. 28

Lesson 9/pages 25-27

1. 1 box
2. $2,643
3. 141 pounds
4. 21,870 cookies
5. 21 square yards
6. $1,142
7. $1,908
8. $3,750

Life Skill/pages 28-29

1. $69
2. $68.75

Unit 1 Posttest/pages 30-31

1. 80
2. 500
3. 1,000
4. 8
5. Forty-six
6. Twenty-five thousand, one hundred eighty-three
7. 378
8. 9,604
9. 2,780
10. 35,000
11. 3,000,000
12. 8,679
13. 3,053
14. 226
15. 139
16. 338
17. 8,187
18. 10,968
19. 45
20. 1,406*
21. 1,368
22. 6,120
23. 468
24. 88
25. $6,800
26. 640 people
27. 45 letters
28. $30

Unit 2 Pretest/pages 32-33

1. Six tenths
2. Twenty-two hundredths
3. Six hundred twenty-three thousandths
4. Seven and nine tenths
5. Twenty-four and six tenths
6. Two and three hundred twenty-four thousandths
7. 0.3
8. 0.214
9. 9
10. 8.61
11. 82.442
12. 146.167
13. 57.51
14. 20.442
15. 0.0192
16. 2.9169
17. 0.0336
18. 21.95856
19. 720
20. 0.072
21. 22
22. 25
23. 0.38 inches
24. $1,029.90

Lesson 10/pages 34-35

1. Eight tenths
2. Nine hundredths
3. Three hundred twenty-seven thousandths
4. Nine and seven tenths
5. Sixty-eight and two thousand five hundred forty-three ten-thousandths
6. 0.6
7. 0.142
8. 1,510
9. 8.08
10. 219.6

Lesson 11/pages 36-37

1. 29.052
2. 1.696
3. 119.675
4. 54.48
5. 39.03
6. 0.608
7. 51.68
8. 30.718
9. 32.068
10. 28.2

Lesson 12/pages 38-39

1. 0.24
2. 40.32
3. 0.027
4. 57.96
5. 0.0096
6. 2.2824
7. 0.0006
8. 2.21733
9. $8.94
10. 860.475 miles

Lesson 13/pages 40-41

1. 1,800
2. 0.18
3. 0.018
4. 0.018
5. 3.420
6. 23.023
7. 22
8. 5.763
9. 1.994
10. 8.171

Life Skill/pages 42-43

1. 23.864 miles per gallon
2. 32.42 miles per gallon

Lesson 14/pages 44-45

1. $521.90
2. 376.8 gallons
3. $2,842.31
4. 103.495 inches
5. $24

Unit 2 Posttest/pages 46-47

1. Three tenths
2. Seventy-two hundredths
3. Three hundred sixty-three thousandths
4. Seven and twenty-nine hundredths
5. Sixty-two and forty-six hundredths
6. Four and nine hundred twenty-five thousandths
7. 0.2
8. 0.624
9. 12
10. 7.87
11. 33.402
12. 296.6
13. 18.51
14. 11.01
15. 11.842
16. 31.7
17. 0.0184
18. 1.9208
19. 0.1323
20. 5.88336
21. 23

22. 56
23. 94.44
24. 13.75
25. 0.54 inches
26. 45 pies

Unit 3 Pretest/pages 48-49

1. $\frac{1}{3}$
2. $\frac{1}{4}$
3. $\frac{1}{5}$
4. $\frac{3}{4}$
5. $6\frac{1}{4}$
6. $6\frac{1}{5}$
7. $11\frac{2}{9}$
8. $13\frac{2}{7}$
9. $\frac{25}{9}$
10. $\frac{17}{10}$
11. $\frac{10}{3}$
12. $\frac{17}{6}$
13. $\frac{3}{8}, \frac{4}{8}$
14. $\frac{10}{16}, \frac{9}{16}$
15. $>$
16. $=$
17. $<$
18. $<$
19. $\frac{8}{21}$
20. $\frac{3}{10}$
21. $\frac{5}{6}$
22. $\frac{2}{7}$
23. $1\frac{1}{2}$
24. $2\frac{5}{7}$
25. 2
26. $1\frac{7}{20}$
27. $1\frac{1}{12}$
28. $\frac{3}{4}$
29. $\frac{1}{6}$
30. $\frac{1}{6}$
31. $5\frac{5}{8}$
32. $1\frac{44}{45}$
33. $2\frac{1}{8}$ yards
34. $2\frac{3}{8}$ hours

Lesson 15/pages 50-52

1. $\frac{2}{3}$
2. $\frac{2}{11}$
3. $\frac{1}{4}$
4. $\frac{2}{3}$
5. $\frac{5}{6}$
6. $\frac{1}{2}$
7. $\frac{1}{5}$
8. $\frac{4}{5}$
9. $5\frac{1}{2}$
10. $1\frac{5}{8}$
11. $3\frac{3}{8}$
12. $6\frac{1}{2}$
13. $4\frac{2}{3}$
14. $7\frac{1}{9}$
15. $6\frac{1}{4}$
16. $5\frac{1}{9}$
17. $\frac{35}{16}$
18. $\frac{7}{4}$
19. $\frac{14}{11}$
20. $\frac{7}{3}$
21. $\frac{11}{3}$
22. $\frac{5}{4}$
23. $\frac{22}{7}$
24. $\frac{13}{5}$

Lesson 16/pages 53-54

1. 8
2. 16
3. 6
4. 45
5. 14
6. 16
7. 15
8. 15
9. 34
10. 36
11. 72
12. 63
13. 26
14. 35
15. 33
16. 35
17. 24
18. 42
19. 105
20. 40

Lesson 17/pages 55-56

1. 36
2. 25
3. 6
4. 15
5. 42
6. 16
7. 14
8. 24
9. 36
10. 12
11. 20
12. 22
13. 28
14. 36
15. 28
16. 18
17. 45
18. 36
19. 30
20. 24
21. 35

Lesson 18/pages 57-58

1. $>$
2. $<$
3. $>$
4. $=$
5. $>$
6. $<$
7. $<$
8. $=$
9. $=$
10. $>$
11. $>$
12. $>$
13. $>$
14. $>$
15. $<$
16. $\frac{1}{2}, \frac{1}{4}, \frac{1}{6}$
17. $\frac{3}{5}, \frac{2}{5}, \frac{1}{5}$
18. $\frac{5}{6}, \frac{4}{5}, \frac{2}{3}, \frac{1}{2}$
19. $\frac{21}{24}, \frac{6}{7}, \frac{5}{6}, \frac{5}{7}$
20. $\frac{1}{2}, \frac{1}{3}, \frac{1}{5}, \frac{1}{4}$
21. $\frac{7}{8}, \frac{3}{4}, \frac{3}{8}, \frac{1}{4}$
22. $\frac{2}{5}, \frac{3}{7}, \frac{4}{9}, \frac{5}{11}$
23. $\frac{2}{3}, \frac{3}{4}, \frac{4}{5}, \frac{5}{6}$
24. $\frac{1}{5}, \frac{1}{2}, \frac{2}{3}, \frac{3}{4}$
25. $\frac{1}{8}, \frac{2}{4}, \frac{4}{5}, \frac{9}{10}$

Lesson 19/pages 59-61

1. $\frac{4}{21}$
2. $\frac{1}{2}$
3. $\frac{4}{33}$
4. $\frac{1}{15}$

5. $\frac{1}{18}$
6. $\frac{1}{42}$
7. $\frac{7}{60}$
8. $\frac{1}{30}$
9. $\frac{1}{3}$
10. $\frac{1}{12}$
11. $\frac{7}{16}$
12. $\frac{10}{63}$
13. $\frac{5}{8}$
14. $\frac{6}{11}$
15. $\frac{3}{16}$
16. $\frac{1}{16}$
17. $\frac{2}{27}$
18. $\frac{3}{50}$
19. $1\frac{2}{7}$
20. $\frac{3}{8}$
21. $\frac{3}{4}$
22. $\frac{5}{6}$
23. $1\frac{2}{3}$
24. $2\frac{1}{2}$
25. 5 loaves
26. $47\frac{1}{2}$ pounds
27. $5\frac{5}{8}$ cups

Lesson 20/pages 62-63

1. $\frac{9}{11}$
2. $\frac{9}{10}$
3. $1\frac{1}{4}$
4. 1
5. $\frac{1}{6}$
6. $\frac{6}{25}$
7. $\frac{7}{11}$
8. $\frac{1}{8}$
9. $4\frac{4}{5}$
10. $1\frac{5}{18}$
11. $\frac{1}{20}$
12. $6\frac{23}{45}$
13. $1\frac{1}{8}$
14. $4\frac{55}{63}$
15. $4\frac{19}{45}$
16. $2\frac{1}{7}$

Life Skill/page 64

Morgan	$5\frac{1}{4}$
Phyllis	$3\frac{1}{2}$
Sarah	$4\frac{3}{4}$
Norman	$3\frac{1}{2}$
Lucy	4
Gabriel	$2\frac{1}{4}$

Lesson 21/pages 65-66

1. $\frac{15}{16}$ inches
2. 4 hours
3. $7\frac{1}{4}$ yards

Unit 3 Posttest/pages 67-68

1. $\frac{1}{2}$
2. $\frac{1}{5}$
3. $\frac{1}{4}$
4. $\frac{1}{2}$
5. $8\frac{1}{3}$
6. $4\frac{3}{7}$
7. $10\frac{1}{10}$
8. $11\frac{5}{8}$
9. $\frac{20}{9}$
10. $\frac{37}{10}$
11. $\frac{19}{6}$
12. $\frac{19}{7}$
13. $\frac{5}{8}, \frac{4}{8}$
14. $\frac{9}{24}, \frac{18}{24}$
15. $<$
16. $=$
17. $<$
18. $>$
19. $\frac{2}{7}$
20. $\frac{12}{35}$
21. $\frac{2}{3}$
22. $\frac{8}{45}$
23. 7
24. $2\frac{39}{40}$
25. $11\frac{1}{3}$
26. $\frac{17}{20}$
27. $1\frac{1}{18}$
28. $1\frac{1}{4}$
29. $\frac{13}{30}$
30. $\frac{1}{18}$
31. $9\frac{8}{15}$
32. $3\frac{22}{27}$
33. 12 miles
34. 15 hours

Unit 4 Pretest/pages 69-70

1. 0.8%
2. 38%
3. 115%
4. 50%
5. $16\frac{2}{3}\%$
6. 10%
7. 0.15
8. 1.5
9. 0.035
10. $\frac{1}{5}$
11. $1\frac{1}{10}$
12. $\frac{1}{8}$
13. $2\frac{2}{5}$
14. $\frac{3}{7}$
15. $1\frac{3}{7}$
16. $\frac{3}{1}$
17. $\frac{7}{13}$
18. $\frac{4}{5}$
19. $\frac{4 \text{ pints}}{1 \text{ pint}}$
20. $\frac{1 \text{ quart}}{5 \text{ pints}}$
21. 35
22. 8
23. 25%
24. 12.5%
25. 50
26. 36
27. 200%
28. 50%
29. $33\frac{1}{3}\%$
30. 24 parks

Lesson 22/pages 71-72

1. 75 squares should be shaded
2. 75 squares should be shaded
3. 75 squares should be shaded
4. 75 squares should be shaded
5. 4; 75

Lesson 23/pages 73-75

1. 0.45
2. 1.98
3. 0.02
4. 0.007
5. 90%
6. 900%
7. 0.3%

8. 65%
9. 0.06
10. 0.8%
11. 0.15
12. 65%
13. 0.35
14. 80%
15. 0.80

Lesson 24/pages 76-78

1. $\frac{3}{5}$
2. $\frac{1}{4}$
3. $\frac{11}{150}$
4. $\frac{1}{2}$
5. $\frac{31}{200}$
6. $1\frac{4}{5}$
7. $\frac{2}{3}$
8. $\frac{111}{500}$
9. $\frac{3}{25}$
10. $\frac{19}{20}$
11. $\frac{7}{20}$
12. $\frac{2}{5}$
13. $\frac{3}{20}$
14. $\frac{9}{100}$
15. $\frac{3}{25}$
16. $\frac{2}{1}$
17. 40%
18. 290%
19. $2\frac{6}{7}\%$
20. $71\frac{3}{7}\%$
21. 80%
22. 20%
23. $66\frac{2}{3}\%$
24. $12\frac{1}{2}\%$
25. 50%
26. 90%
27. 40%
28. 24%
29. $16\frac{2}{3}\%$
30. $11\frac{1}{9}\%$
31. $14\frac{2}{7}\%$
32. $66\frac{2}{3}\%$

Life Skill/page 79

VCR	$12\frac{1}{2}\%$
Washer	25%
Dryer	40%
Toaster	50%
Camera	20%
Computer	$16\frac{2}{3}\%$
Mixer	$66\frac{2}{3}\%$
Electric Knife	75%

Lesson 25/page 80

1.

300	75
400	100

2.

1.5	1.5
100	100

3.

6	$33\frac{1}{3}$
18	100

4.

133	20
665	100

Lesson 26/pages 81-85

1. 297
2. 550
3. 49
4. 16
5. 40%
6. 25%
7. 2%
8. 400%
9. 665
10. 38
11. 24
12. 70
13. 400
14. 800
15. 50
16. 50

Lesson 27/pages 86-87

1. 22%
2. 45%
3. 41%
4. 45%
5. 50%

Lesson 28/pages 88-89

1. 105 miles
2. 1,140 pounds
3. $78

Lesson 29/pages 90-91

1. $27\frac{7}{9}\%$
2. $22\frac{2}{9}\%$
3. 21%
4. 27%

Lesson 30/pages 92-93

1. $4.68
2. $312.50
3. $1,923.08

Lesson 31/pages 94-95

1. $14,500; 0.09; $1,305; 1 year
2. $1,600; 0.055; $176; 2 years
3. $288
4. $840
5. $135

Lesson 32/pages 96-97

1. $\frac{15}{4}$
2. $\frac{3}{5}$
3. $\frac{9}{7}$
4. $\frac{8}{3}$
5. $\frac{7}{12}$
6. $\frac{3}{5}$
7. $\frac{\text{2 quarts}}{\text{1 quart}}$
8. $\frac{\text{6 pints}}{\text{5 pints}}$
9. 18 miles:1 gallon
10. 1 inch:10 miles
11. 3 cups:1 person
12. 5 yards:1 gown
13. 2:1
14. 21:1
15. 3:5
16. 3:1
17. 10:1
18. 3:1

Lesson 33/pages 98-99

1. 55
2. 2
3. 28
4. $\frac{11}{18}$
5. (1) 4:1
   (2) 1:5
   (3) 4:5
6. (1) 2:3
   (2) 3:2
   (3) 2:5
   (4) 3:5

Lesson 34/pages 100-101

1. 32 ounces

2. 731.25 square feet
3. 18 hours
4. 8.5 miles
5. $3.30

Unit 4 Posttest/pages 102-103

1. 36.2%
2. 63%
3. 145%
4. 0.17
5. 1.45
6. 0.045
7. $2\frac{1}{12}$%
8. $5\frac{5}{39}$%
9. $62\frac{1}{2}$%
10. $\frac{2}{5}$
11. $1\frac{7}{50}$
12. $\frac{111}{200}$
13. 49
14. 3.2
15. 21
16. $1\frac{5}{7}$
17. $1\frac{2}{3}$
18. $2\frac{5}{7}$
19. $\frac{10 \text{ pints}}{3 \text{ pints}}$
20. $\frac{6 \text{ quarts}}{5 \text{ gallons}}$
21. 4 cups
22. 5:1
23. $4,750
24. 15%
25. $16\frac{2}{3}$%
26. 20%

Unit 5 Pretest/pages 104-106

1. Montreal
2. Albuquerque
3. 24 in.
4. 15 in.
5. MS Word
6. AutoCad
7. 8 releases
8. 2 releases
9.

Company	Mean	Median	Mode	Range
X	165.625	160	160	15
Y	161.625	160	160	25
Z	163.50	160	150, 160, 175	30

(1) Median
(2) Company X

Lesson 35/pages 107-114

1. Sales and gross receipts
2. Motor vehicle and operators' licenses
3. Sales and gross receipts
4. Other
5. 6 feet
6. 1.5 feet
7. 5 feet
8. 2 frogs
9. 58%
10. 45%
11. Medical assistants
12. Medical assistants, home health aides, radiologic technologists
13. Electrical and electronics engineers, information clerks
14. 86
15. 22
16. 24
17. 1978
18. 1988

Life Skill/page 115

1. B
2. C
3. A

Lesson 36/pages 116-118

1.

Mean	Median	Mode	Range
34	33	None	14
35.2	28	None	32

2.

Mean	Median	Mode	Range
15	15	14,16	2
14.25	15	15	6
13.875	13	11,13	8

Company X because it has a higher mean and lower range.

3.

Mean	Median	Mode	Range
35,000	27,500	25,000	35,000
35,000	35,000	35,000	0

(1) Median and mode
(2) Median and mode

Lesson 37/pages 119-121

1. 90

Sum of Scores
790
90
880

2. 80

Sum of Scores
70
76
86
88
80
400

3.

Amount Spent
7,500
6,000
4,500
3,600
3,900
3,000
1,500
30,000

Unit 5 Posttest/pages 122-123

1. Tokyo
2. New York, Osaka
3. Mexico City, São Paulo
4. 16,000
5. New York
6. Bombay
7. 175
8. Seoul
9. 14.35
10. 14, 16
11. 14
12. 8
13. Mean

Unit 6 Pretest/pages 124-126

1. Ray *AB*
2. Square *ABCD*
3. Parallelogram *ABCD*
4. Trapezoid *ABCD*
5. $\angle CBE$, $\angle BED$, $\angle ABG$, or $\angle FEH$
6. $\angle GBC$, $\angle ABE$, $\angle BEF$, or $\angle DEH$
7. $\angle ABG$ and $\angle BED$, $\angle GBC$ and $\angle BEF$, $\angle ABE$ and $\angle DEH$, $\angle CBE$ and $\angle FEH$
8. $\angle GBA$ and $\angle CBE$, $\angle GBC$ and $\angle ABE$, $\angle BEF$ and $\angle DEH$, $\angle BED$ and $\angle FEH$
9. $\angle ABG$ and $\angle FEH$, $\angle GBC$ and $\angle HED$
10. $\angle ABE$ and $\angle FEB$, $\angle CBE$ and $\angle DEB$
11. It has a 90° angle.
12. All angles measure less than 90°.
13. P = 31 ft; A = 44 sq ft

14. P = 38 in.; A = 68 sq in.
15. 240 mm
16. 36 ft
17. 13 in.
18. 5 cm

Lesson 38/pages 127-130

1. Ray *AB*
2. Line segment *CD*
3. ∠*C*, ∠*DCE*, or ∠*ECD*
4. Line *DF*
5. ∠*A* or ∠*CAE*
6. Ray *ML*
Answers will vary.
7. 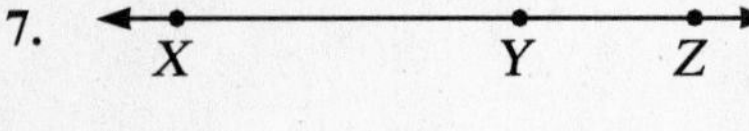

XY, XZ, YZ, ZX
8. 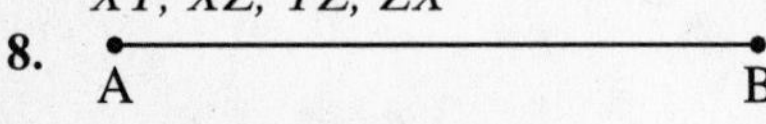

A line segment has two endpoints.
9. 

A ray has only one endpoint, and a line has no endpoints.

Lesson 39/pages 131-135

1. No
2. Yes
3. ∠*AOB* and ∠*BOD*; ∠*DOC* and ∠*COA*
4. ∠*DOC* and ∠*AOB*; ∠*DOA* and ∠*COB*
5. ∠2, ∠4, ∠6, or ∠8
6. ∠1, ∠3, ∠5, or ∠7
7. ∠4 and ∠6; ∠3 and ∠5
8. ∠1 and ∠7; ∠2 and ∠8
9. ∠1 and ∠5; ∠4 and ∠8; ∠2 and ∠6; ∠3 and ∠7
10. ∠1 and ∠3; ∠2 and ∠4; ∠5 and ∠7; ∠6 and ∠8

Lesson 40/pages 136-138

1. △*DEF*
2. △*ABC*
3. △*LMN*

Life Skill/page 139

1. 80%; 42%; 20%
2. A

Lesson 41/pages 140-142

1. 40 cm
2. 15 in.
3. 22 ft
4. 64 m
5. 56 cm
6. 72 in.
7. 88 ft
8. 46 in.
9. 94 ft.

Lesson 42/pages 143-145

1. 90 cm^2
2. 12 ft^2
3. 81 $in.^2$
4. 735 cm^2
5. 33 $in.^2$
6. 220 ft^2
7. 126 cm^2
8. 96 m^2
9. 140 ft^2

Lesson 43/pages 146-149

1. ∠*A*~∠*D*, ∠*B*~∠*E*, ∠*C*~∠*F*; $\overline{AB}\sim\overline{DE}$, $\overline{BC}\sim\overline{EF}$, $\overline{AC}\sim\overline{DF}$
2. 8
3. ∠*G*~∠*J*, ∠*H*~∠*K*, ∠*I*~∠*L*; $\overline{GH}\sim\overline{JK}$, $\overline{HI}\sim\overline{KL}$, $\overline{GI}\sim\overline{JL}$
4. 4
5. ∠*M*~∠*P*, ∠*N*~∠*Q*, ∠*O*~∠*R*; $\overline{MN}\sim\overline{PQ}$, $\overline{NO}\sim\overline{QR}$, $\overline{MO}\sim\overline{PR}$
6. 20
7. 90 mm
8. 25 in.

Lesson 44/pages 150-153

1.

a	*b*	a^2	b^2	c^2	*c*
3	4	9	16	25	5
6	8	36	64	100	10
5	12	25	144	169	13
12	5	144	25	169	13
20	21	400	441	841	29

2. 7.07
3. 13
4. 17
5. 14.14
6. No
7. Yes
8. Yes
9. No

Lesson 45/pages 154-155

1. 23 feet
2. 60°; 80°
3. 960 feet

Unit 6 Posttest/pages 156-158

1. Parallelogram *ABCD*
2. Line *AB*
3. Angle *B*, Angle *ABC*, or Angle *CBA*
4. Rectangle *EFGH*
5. ∠2, ∠4, ∠5, ∠7, ∠10, or ∠11
6. ∠1, ∠3, ∠6, ∠8, ∠9, or ∠12
7. ∠3 and ∠6, ∠4 and ∠5, ∠9 and ∠12, ∠10 and ∠11
8. ∠2 and ∠7, ∠1 and ∠8
9. Two sides are equal.
10. No sides are equal.
11. P: 152; A: 1,068
12. P: 100; A: 480
13. 20 cm
14. 30 ft
15. 50
16. 11.3

Unit 7 Pretest/pages 159-160

1. $\overline{CA}$, $\overline{CD}$, or $\overline{CB}$
2. $\overline{AB}$ or $\overline{BA}$
3. *C*
4. $\overline{AB}$ or $\overline{EF}$
5. C = 37.68 ft; A = 113.04 ft^2
6. C = 31.4 in.; A = 78.5 $in.^2$
7. C = 18.84 cm; A = 28.26 cm^2
8. C = 25.12 ft; A = 50.24 ft^2
9. 31.4 ft
10. 12.56 $in.^2$

Lesson 46/pages 161-163

1.

Radius	Diameter	Circumference	Circumference/Diameter
2.5	5	15.7	3.14
2	4	12.56	3.14
5	10	31.4	3.14
1	2	6.28	3.14

2. 31.4 in.
3. 25.12 cm
4. 37.68 yd
5. 37.68 in.
6. 47.1 yd
7. 75.36 m
8. 21.98 in.
9. 37.68 ft

Life Skill/pages 164-165

1. 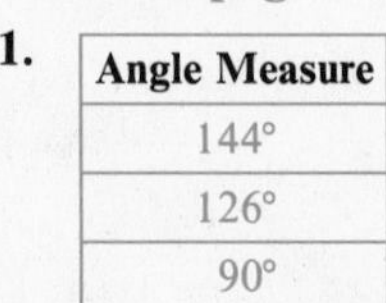

Angle Measure
144°
126°
90°

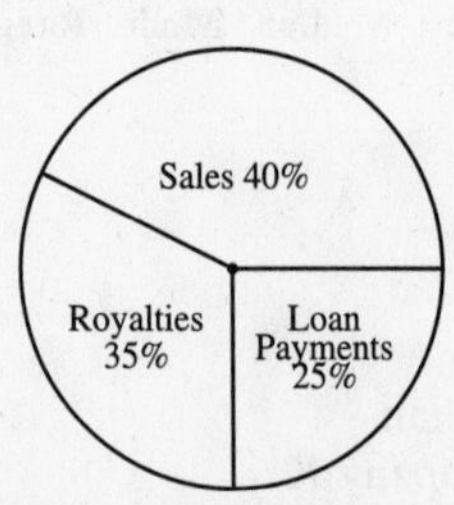

2. \$2,000,000
3. \$1,750,000
4. \$1,250,000

Lesson 47/pages 166-168

1. 28.26 ft^2
2. 19.625 in^2
3. 60
4. 84
5. 115.25 ft^2
6. 1.57 ft^2
7. 7.065 yd^2
8. 7 bags
9. 113.04 ft^2

Lesson 48/pages 169-170

1. 157 ft^2
2. 123.84 in^2

Unit 7 Posttest/pages 171-172

1. $\overline{DC}$, $\overline{DE}$, or $\overline{DF}$
2. $\overline{CF}$ or $\overline{FC}$
3. D
4. $\overline{AB}$ or $\overline{CF}$
5. C = 6.28 ft;
   A = 3.14 ft^2
6. C = 25.12 in.;
   A = 50.24 $in.^2$
7. C = 31.4 yd;
   A = 78.5 yd^2
8. C = 25.12 ft;
   A = 50.24 ft^2
9. 8¢
10. 50.24 in.

Unit 8 Pretest/pages 173-174

1. V = 1,250 $in.^3$; SA = 850 $in.^2$
2. V = 120 cm^3; SA = 184 cm^2
3. V = 64 ft^3; SA = 144 ft^2
4. V = 254.34 ft^3; SA = 226.08 ft^2
5. V = 2,260.8 ft^3; SA = 979.68 ft^2
6. V = 37.7 ft^3; SA = 75.36 ft^2
7. V = 33.49 ft^3; SA = 50.24 ft^2
8. V = 113.04 ft^3; SA = 113.04 ft^2
9. V = 2,080 $in.^3$; SA = 888 $in.^2$

Lesson 49/pages 175-178

1. 391.44 in^2
2. 376 ft^2
3. 215 m^2
4. 318 ft^2

Lesson 50/pages 179-181

1. 12,500 m^3
2. 7,581.6 $in.^3$
3. 80 yd^3
4. 576 ft^3
5. 160 $in.^2$

Lesson 51/pages 182-185

1. V = 2,512 ft^3, LA = 502.4 ft^2, SA = 1,130.4 ft^2
2. V = 579.4 cm^3, LA = 373.8 cm^2, SA = 434.15 cm^2
3. V = 452.16 cm^3, LA = 252.5 cm^2, SA = 365.5 cm^2
4. V = 471 m^3, LA = 293.6 m^2, SA = 372.1 m^2
5. 21,703.68 bushels
6. 9¢ per $in.^3$

Life Skill/pages 186-187

1. \$2.40

Lesson 52/pages 188-189

1. V = 0.523 $in.^3$; SA = 3.14 $in.^2$
2. V = 8.178 $in.^3$; SA = 19.625 $in.^2$
3. V = 381.51 $in.^3$; SA = 254.34 $in.^2$
4. V = 260,579,713,200 cubic miles
   SA = 197,259,434.6 square miles
5. V = 5,273,994,240 cubic miles
   SA = 14,649,984 square miles
6. V = 337,535,631,400,000,000 cubic miles
   SA = 2,343,997,440,000 square miles

Lesson 53/pages 190-191

1. 5,652 bushels
2. 1186.72 square feet
3. 8 pints

Unit 8 Posttest/pages 192-193

1. V = 1,200 cm^3; SA = 820 cm^2
2. V = 81 ft^3; SA = 184 ft^2
3. V = 37.68 yd^3; SA = 75.36 yd^2
4. V = 2,560 m^3; SA = 1,280 m^2
5. V = 4,536 $in.^3$; SA = 1,757 $in.^2$
6. V = 5861.34 $in.^3$; SA = 2009.6 $in.^2$
7. 3768 $in.^3$

Unit 9 Pretest/pages 194-195

1. >
2. <
3. >
4. <
5. =
6. <
7. 2
8. −27
9. 23
10. 5
11. 0
12. $6\frac{1}{5}$
13. $-2\frac{1}{4}$
14. −29.3
15. 36
16. −56
17. −48
18. 65
19. $-\frac{2}{3}$
20. −1
21. −3
22. −126
23. 840
24. 715
25. Undefined
26. −40
27. −22
28. −8 feet

Lesson 54/pages 196-198

1. True
2. True
3. True
4. True
5. True
6. True
7. −10
8. 1
9. 13
10. −18
11. −1
12. −33
13. −5, −2, 0, 1
14. −2, 3, 4, 5
15. 4 feet above normal
16. 6 feet below normal
17. 10 feet
18. 7 feet
19. Dead Sea
20. Mt. McKinley
21. Mariana Trench
22. Death Valley

Lesson 55/pages 199-201

1. 4
2. −4
3. −2
4. 2
5. 7
6. −7
7. 1
8. 8
9. −6
10. −8
11. −13
12. −29
13. −86
14. $-33\frac{3}{4}$
15. 29.5
16. −10.1

17. -31
18. 0
19. -15
20. -11
21. -11
22. 10
23. 17
24. $20\frac{3}{4}$
25. 9
26. 17
27. -96.6
28. -9
29. 296
30. 0
31. 65°
32. 50°

Lesson 56/pages 202-204

1. -21
2. -30
3. 56
4. 36
5. -81
6. -48
7. 195
8. $\frac{5}{4}$
9. -2
10. $-\frac{1}{8}$
11. 36
12. 1
13. -1
14. -36
15. -13
16. -3
17. -13
18. 2
19. $-1{,}440$
20. 20,580
21. 0
22. 0
23. 1,650
24. $-2\frac{1}{2}$
25. Undefined
26. 0

Lesson 57/pages 205-206

1. -26
2. 19
3. 27
4. 117
5. $-2\frac{1}{2}$
6. 34
7. 2
8. 2
9. -5
10. 1
11. $\frac{1}{7}$
12. 1
13. -42
14. 301
15. $\frac{16}{81}$
16. -0.96
17. -0.64
18. 4
19. 0
20. $-3\frac{1}{5}$

Lesson 58/pages 207-209

1. $65.25
2. $40.80
3. 3 yard gain
4. $27,000
5. -7 under par
6. 50°
7. 60°
8. $170

Life Skill/pages 210-211

Yes, the adjusted balance for both is $3,168.74.

Unit 9 Posttest/pages 212

1. $>$
2. $<$
3. $<$
4. $>$
5. $<$
6. $>$
7. 10
8. -25
9. -65
10. 13.8
11. 6
12. -11
13. -39
14. -29
15. $-2\frac{1}{5}$
16. 126
17. -63
18. 294
19. -3
20. -7
21. $-\frac{1}{24}$
22. -144
23. -200
24. -208
25. Undefined
26. 16
27. $-\frac{2}{3}$
28. $240

Unit 10 Pretest/pages 213-214

1. $x + 6$
2. $x - 25$
3. $5\left(\frac{x}{10}\right)$
4. $(x + 1) - 25$
5. 100
6. 40
7. 24 ft
8. 3 ft
9. $19a - 9$
10. $16ac + ab$
11. $-4x - 11y$
12. $-5v^2 + 2v$
13. $2\frac{1}{2}$
14. -3
15. -1
16. $\frac{1}{2}$
17. 2.6
18. 11
19. 9
20. 14
21. -4
22. $5\frac{1}{2}$
23. 91
24. $300

Lesson 59/pages 215-216

1. $x + 2$
2. $6x$
3. $x - 10$
4. $\frac{x}{5}$
5. x^2
6. $\frac{x}{2}$
7. $x + 4$
8. $x + (-7)$
9. $x - 12$
10. $x - 12$
11. $2x + 10$
12. $4\left(\frac{x}{3}\right)$
13. $x - (0.15x)$
14. $x^2 + 5$
15. $\frac{1}{5}(x + 1)$
16. $50 - x$

Lesson 60/pages 217-219

1. 52
2. 26
3. 10
4. 122
5. 35
6. 300
7. 38
8. 192.5
9. $A = s^2$; 39.0625
10. $P = 4s$; 25 in.^{2}
11. $A = lw$; 34.4375 yd^2

12. $P = 2l + 2w$; 24 yd
13. $M = \frac{(A + B + C + D)}{4}$; 75
14. $I = prt$; \$144
15. $V = lwh$; 109 in.^{3}
16. $\frac{d}{r} = t$; 5 hours
17. $c = p + pr$; \$26.44
18. $s = p - pd$; \$80.75

Lesson 61/pages 220-222

1. $7x$
2. $2y$
3. $-2a$
4. $-4abc$
5. $4x^2y$
6. 0
7. $-x + 7y$
8. $6x - 6y$
9. $-x^2 + 17x$
10. $-14m^2 - 22m$
11. $-2m^2 + 4m - 6$
12. $8z$
13. $2x + 10$
14. $6x + 3y$
15. $4x + 8y$
16. $4x - 12$
17. $-4x + 20y$
18. $-9x + 6y$
19. $5x + 2$
20. $8x - 3$
21. $6x + 20$
22. $-16x + 22y$
23. \$1,400
24. \$25,800

Lesson 62/pages 223-225

1. 4
2. 5
3. 18
4. 40
5. 6
6. 23
7. 6.2
8. 2.2
9. 2.1
10. 8.4
11. 20
12. 29
13. 14
14. −2
15. −4
16. −7
17. −7
18. −1
19. 0
20. 0
21. 0.4
22. 2
23. −0.4
24. −0.6
25. −1.5
26. −3

Life Skill/pages 226-228

1. \$1.50
2. \$6.34

Lesson 63/pages 229-231

1. 5
2. 7
3. 12
4. 25
5. 16
6. 4
7. 4
8. 3
9. 15
10. −8
11. −6
12. 4
13. 4
14. −1
15. 3
16. 5
17. 8
18. 15
19. 13
20. 9
21. 19
22. 15
23. 18
24. 10
25. 3
26. 3
27. −50
28. 30
29. 4
30. −10
31. 3
32. 5
33. 4
34. −2
35. 3
36. −1
37. −2
38. −14

Lesson 64/pages 232-234

1. 5 ft
2. \$2,500
3. 3 in.
4. 14.14 ft
5. 11 ft
6. 68° F
7. 80
8. 8 ft

Unit 10 Posttest/pages 235-236

1. $x + 7$
2. $x - 15$
3. $\frac{x}{10x}$
4. $4x + 20$
5. \$400
6. 16 yd^2
7. 50 hours
8. 83.25
9. $-t + 2$
10. $14mn - 4m + 6n$
11. $11d + 7c$
12. $-44p^2 + 22p$
13. 4
14. 2
15. −23
16. 19
17. 20
18. 1.3333
19. $\frac{1}{9}$
20. −2
21. 15
22. 5.8
23. \$10,000
24. 25 ft

Unit 11 Pretest/page 237

1. 13
2. 56,58
3. 6 hours
4. 3 oak, 15 pine
5. Quarters = 33, dimes = 70
6. Lisa is 12, Brian is 24

Lesson 65/pages 238-241

1. $x + 6 = 45$
2. $25x = 80$
3. $\frac{64}{x} = 12$
4. $x - \frac{1}{2} = 3$
5. $4x = 24$
6. 21
7. 9
8. 10, 15
9. 12, 18
10. 10, 11
11. 29, 30
12. 18, 20
13. 21, 23
14. 13, 15, 17
15. 2, 4, 6
16. 10, 12, 14
17. 11, 13, 15
18. 2, 4, 6
19. 14, 16, 18

Lesson 66/pages 242-245

1. 275 miles
2. 1,220 miles
3. 9 hours
4. 420 miles

5. David = 4 hours, Matthew = 5 hours
6. About 1.9 miles each way
7. 4 hours
8. 280 miles
9. 100 miles

Lesson 67/pages 246-249

1. \$3,500 at 6%, \$2,500 at 5%
2. Jane is 20; Pat is 10.
3. Adult = 185, Child = 370
4. \$6,000 in stocks; \$4,000 in bonds
5. 15 and 5
6. David is 30; Matthew is 10.
7. \$4,000 at 5%; \$5,500 at 7%

Life Skill/pages 250-251

1. 1.875 hours
2. 6 hours
3. 6 hours
4. 3 hours

Unit 11 Posttest/page 252

1. 6
2. 81, 83, 85
3. 1 hour 20 minutes
4. 13 points for foul shots, 28 points for two-point shots
5. Carl is 24, and Stuart is 32.
6. \$4,500 at 10%; \$3,000 at 8%

Unit 12 Pretest/pages 253-254

1. 4
2. $2p^{12}$
3. $81x^{10}$
4. t^6
5. $2xy^3 + 16xy^4$
6. $20t^3 + 35t^5$
7. $x^2 + 6x + 5$
8. $s^2 - 144$
9. $3x^3 - 7x^2 + 12x - 28$
10. $14 + 13x - x^2$
11. $y^2 - 169$
12. $k^2 + k - 462$
13. $(x + 7)(x + 2)$
14. $(x + 6)(x + 8)$
15. $(x - 4)^2$
16. $(b + 7)(b + 1)$
17. $4(x + 6)(x + 6)$
18. $(x + 5)(x - 5)$
19. $4, -4$
20. -10
21. 3, 2
22. $-4, \frac{3}{2}$
23. 132 games
24. 6 inches

Lesson 68/pages 255-257

1. 27
2. 16
3. 49
4. 64
5. $\frac{1}{4}$
6. 0.000027
7. x^5
8. y^3
9. 2^6
10. $\left(\frac{1}{3}\right)^3$
11. a^6
12. b^{10}
13. 32
14. b^5
15. x^{21}
16. y^{12}
17. a
18. y^2
19. $\frac{1}{x^2}$
20. $\frac{1}{y^3}$
21. a^3
22. x^4
23. $\frac{1}{y^4}$
24. 1
25. x
26. $\frac{1}{c^7}$
27. x^6
28. y^{18}
29. y^{32}
30. 81
31. x^{28}
32. a^{16}
33. x^6y^6
34. a^6b^8
35. 9
36. x^{12}
37. $6x^{22}$
38. x^{14}
39. $3xy^2 + 4x^2y^2$
40. $2x^{10}y^{14}$
41. 100, 121, 144, 169, 196
42. 216, 343

Life Skill/pages 258-259

1. 5×10^3
2. 3.4876×10^4
3. 7.58×10^3
4. 5.8×10^3
5. 1.035×10^6
6. $3.333 \times 10^{(-1)}$
7. 2×10^1
8. 2.56×10^7

Lesson 69/pages 260-261

1. $15x^5$
2. $12x + 8$
3. $-6y^7$
4. $-3x^2 - 27$
5. $8x^3 + 4x$
6. $-30x^4 - 25x^3$
7. $7a^2 - 28a$
8. $32c^5 - 48c^3$
9. $-27y^4 - 18y^3$
10. $6x^2 + 12x + 6$
11. $-27x^2 - 36x - 27$
12. $3x^2y^2 + 3xy^2 + 3y^2$
13. $-y^3$
14. $-8b^5 + 6b^4c$
15. $30a^3x + 10a^2x^2 + 20ax$
16. $-27xy^3 - 18xy^2 - 27x$

Lesson 70/pages 262-263

1. $x^2 + x - 6$
2. $x^2 + 10x + 24$
3. $y^2 - 25$
4. $m^2 - 8m - 20$
5. $a^2b^2 - 2ab - 15$
6. $x^2y^2 - 16xy + 64$
7. $40 + 14x + x^2$
8. $49 - t^2$
9. $6x^2 + 11x - 10$
10. $6m^2 - 12m - 48$
11. $4x^4 + 12x^2 - 27$
12. $9x^2 - 36x + 36$
13. $-2 + p + 15p^2$
14. $2x^3 - 2x^2 - 12x$
15. $-x^2 + 3xy + 4y^2$
16. $3m^2 + mn - 2n^2$

Lesson 71/pages 264-266

1. 4
2. 6
3. 4
4. 6
5. 7
6. 5
7. $2xy$
8. $2x$
9. $5x^2$
10. $8x^2y$
11. $25x^3y$
12. $(2x - y)$
13. $4(2x + 3)$
14. $6(2y - 3)$
15. $3(x + y)$
16. $4(y - 3)$
17. $3x(3 + 4y)$
18. $5x(2x - 3)$
19. $4x^2(3x + 4)$
20. $15x(x^2 - y)$
21. $25a^2b^2(1 - 4a)$
22. $(a + b)(2 + 3a)$

Lesson 72/pages 267-270

1. $-3, -4$

2. $-7, -3$
3. $7, -5$
4. $8, -3$
5. $11, 1$
6. $(x - 5)(x - 1)$
7. $(x + 2)(x + 1)$
8. $(x - 7)(x - 1)$
9. $(x - 5)(x + 3)$
10. $(x + 10)(x + 2)$
11. $(x - 6)(x + 5)$
12. $(x - 7)(x - 3)$
13. $(x - 7)(x + 2)$
14. $(x + 7)^2$
15. $(x - 4)^2$
16. $(x - 9)(x + 11)$
17. $(x - 6)(x - 2)$
18. $(x - 2)(x + 2)$
19. $(x - 3)(x + 3)$
20. $(x - 6)(x + 6)$
21. $(x - 12)(x + 12)$
22. $\left(x - \frac{1}{3}\right)\left(x + \frac{1}{3}\right)$
23. $(x - 0.2)(x + 0.2)$
24. $(x^2 + 9)(x + 3)(x - 3)$
25. $2(x^2 + 3)(x^2 - 3)$
26. $(3x - 5)(x + 2)$
27. $2(2x + 3)(x + 1)$

Lesson 73/pages 271-272

1. $-5, -3$
2. $-3, -4$
3. 3
4. $0, 5$
5. $9, -5$
6. $\frac{6}{5}, 0$
7. -5
8. $4, -4$
9. $0, 7$
10. $-8, -1$
11. $2, 4$
12. 5

Lesson 74/pages 273-275

1. $11, 13$
2. $5, 7$
3. $14, 4$
4. $-7, -9$

Unit 12 Posttest/pages 276-277

1. $4x^2y^3$
2. d^{19}
3. $74{,}088$
4. z^4
5. $x^4 + 2x^5$
6. $2v^3 + 6v^5$
7. $y^2 + 8y - 9$
8. $h^2 - 49$
9. $20x^3 - 12x^2 + 10x - 6$
10. $20 - 21x + x^2$
11. $y^2 - 441$
12. $c^2 + 10cz - 24z^2$
13. $(x + 9)(x - 2)$
14. $(x + 7)(x - 3)$
15. $(x - 4)^2$
16. $(m + 7)(m - 1)$
17. $5(n + 7)(n + 2)$
18. $(x - 11)(x + 11)$
19. $4, -4$
20. $7, 2$
21. $-1, \frac{5}{3}$
22. $\frac{6}{5}, -1$
23. 780

Unit 13 Pretest/pages 278-280

1.-4.

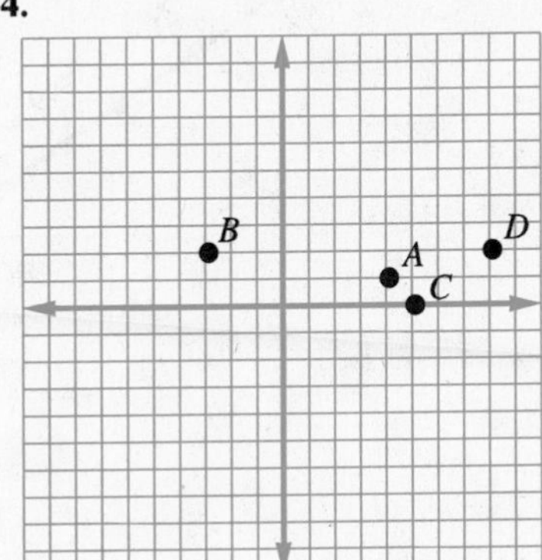

5. 3
6. 5
7. 1
8. 2
9. $6, 0, 4$
10. $5, 1, 11$
11.

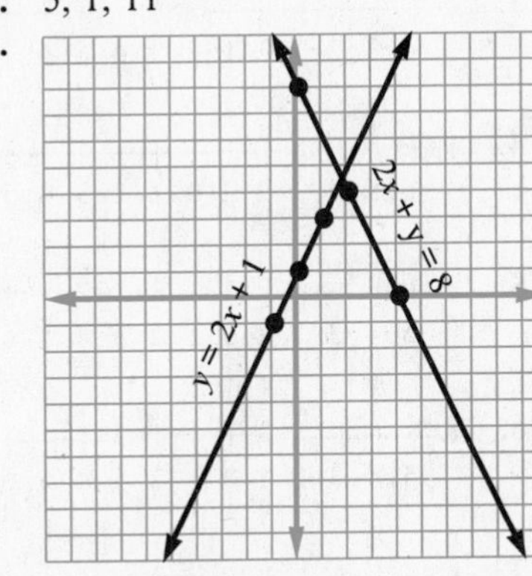

12. -8 -7 -6 -5 -4 -3 -2 -1 0 1 2 3

13. -5 -4 -3 -2 -1 0 1 2 3 4 5 6

14. -5 -4 -3 -2 -1 0 1 2 3 4 5 6

15. -6 -5 -4 -3 -2 -1 0 1 2 3 4 5

16. $x < -10$

-16 -15 -14 -13 -12 -11 -10 -9 -8 -7 -6 -5

17. $x > 3$

-5 -4 -3 -2 -1 0 1 2 3 4 5 6

18. $x \leq -21$

-29 -28 -27 -26 -25 -24 -23 -22 -21 -20

19. $x < 4$

-5 -4 -3 -2 -1 0 1 2 3 4 5 6

20. $x + 15 < 4x;\ x > 5$
21. $3 \geq 0.75 + 0.45x$

Lesson 75/pages 281-284

1.-5.

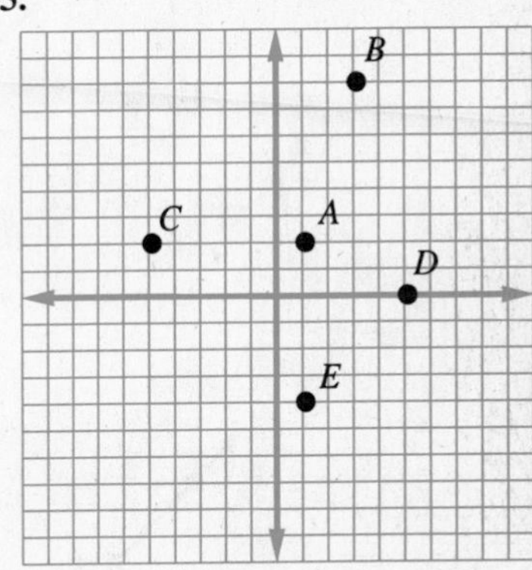

6. 6.3
7. 10
8. 6.4
9. 12.04
10. $\frac{1}{2}$
11. $\frac{3}{2}$
12. $-\frac{1}{2}$
13. Undefined
14. 0
15. Undefined

Lesson 76/pages 285-288

1. $5, 8, 2$
2. $-4, -2, 0$
3. $0, 1, -2$
4. $-2, 3, 5$
5. $9, 21, 13$
6. $1, 5, -1$

7.

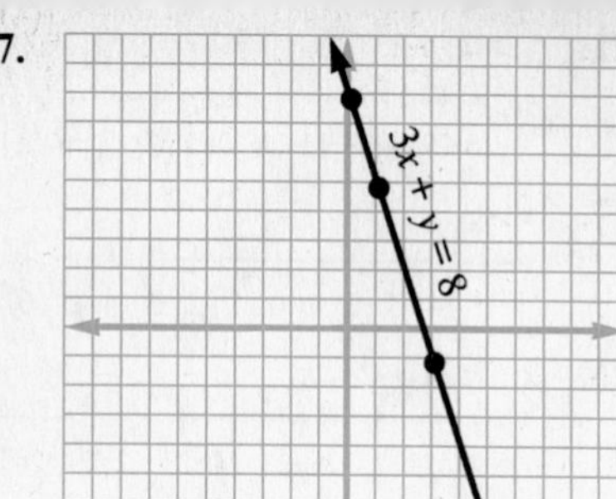

8.

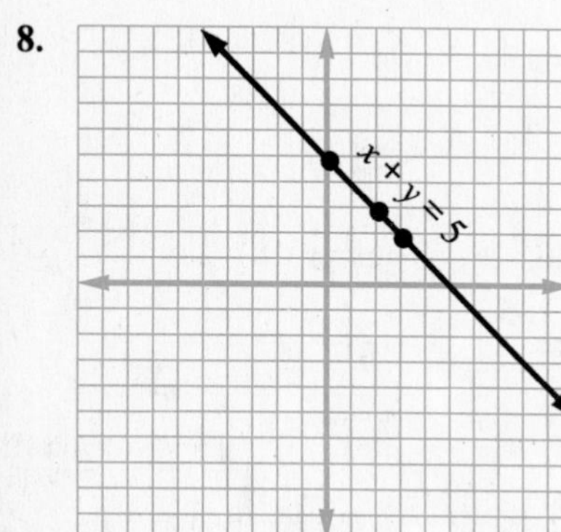

9.

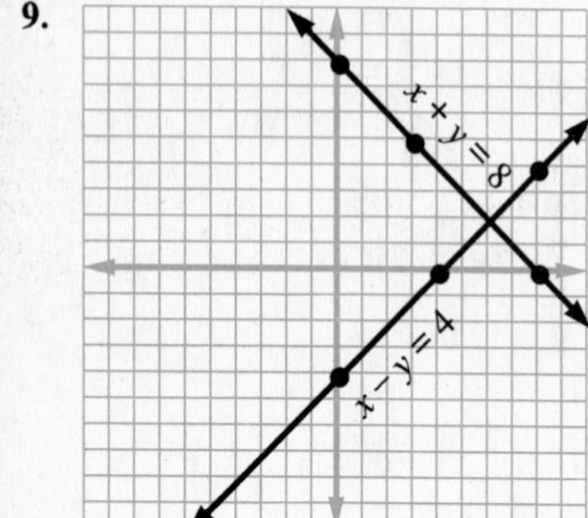

10.

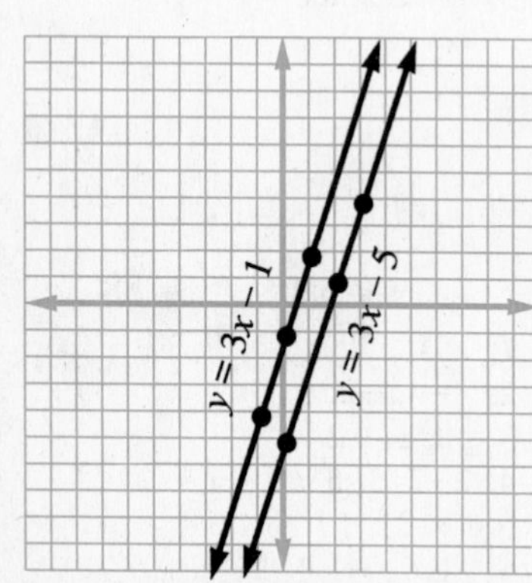

11.

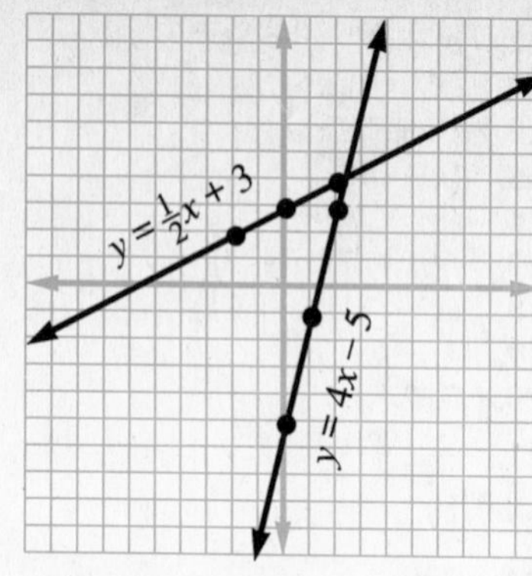

12.

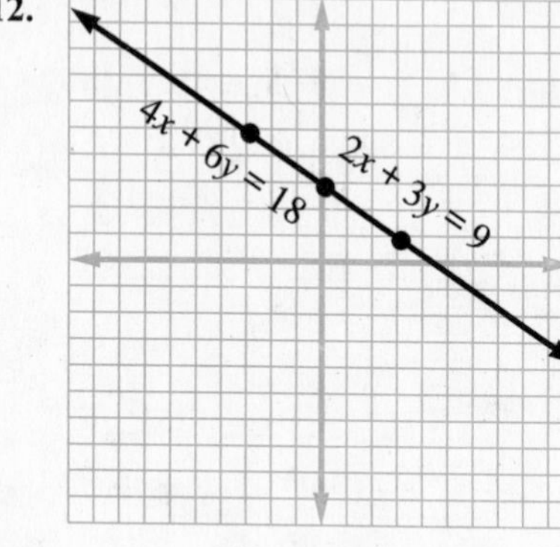

Lesson 77/pages 289-291

1.

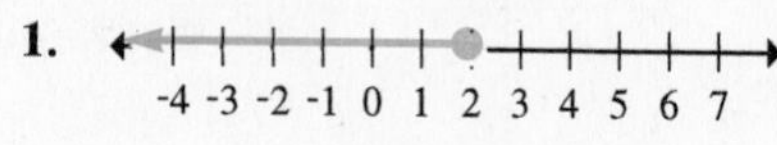

2. -5 -4 -3 -2 -1 0 1 2 3 4 5 6

3. -4 -3 -2 -1 0 1 2 3 4 5 6 7

4.

5.

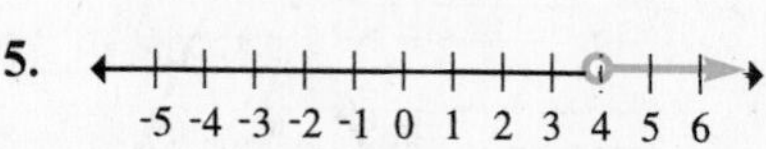

6.

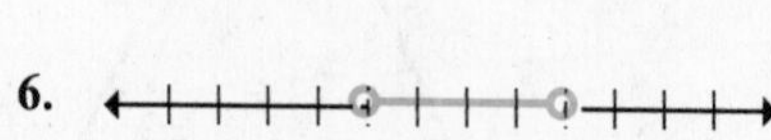

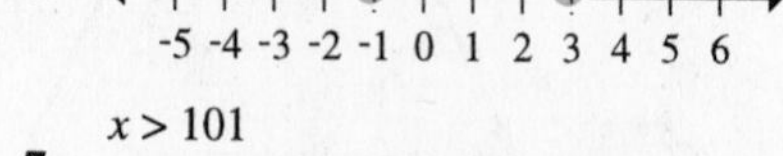

7. $x > 101$

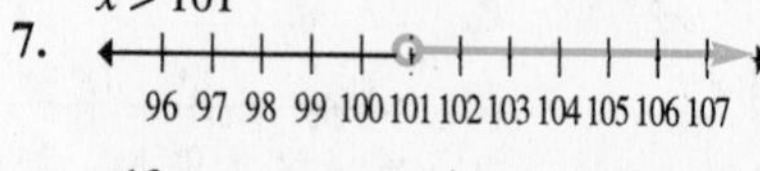

8. $-13 > x$

9. $x > 16$

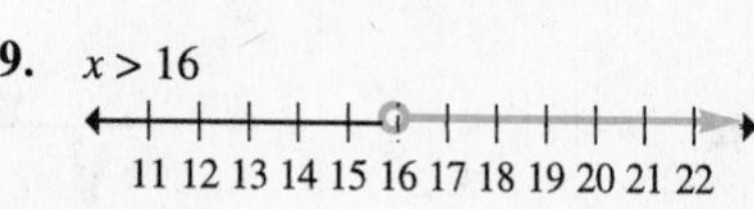

10. $x > 24$

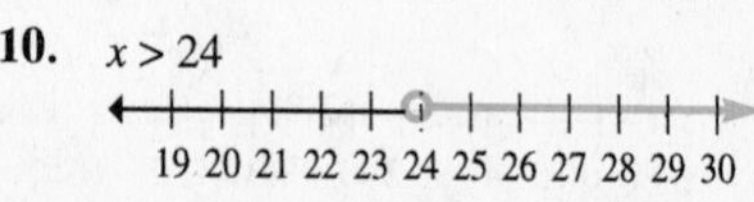

11. $x < 8$

12. $x < 13$

13. $x > -1$

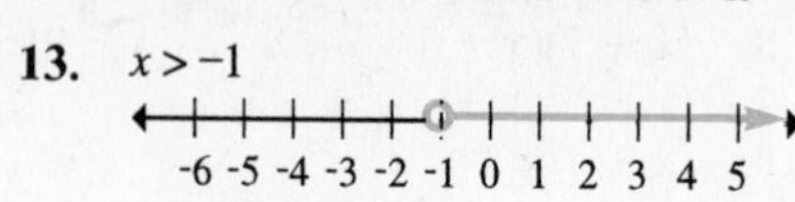

14. $x < -5$

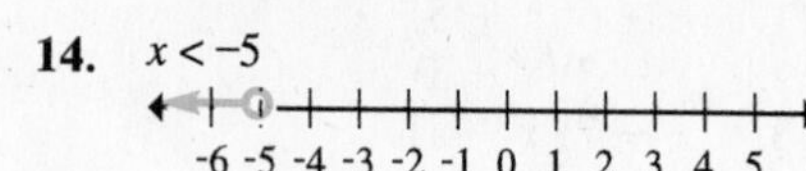

Life Skill/pages 292-293

1. There will be a surplus.
2. There will be a shortage.

Lesson 78/pages 294-295

1. $x < 50$
2. $x > 73$
3. $5.5 < x < 6.5$
4. $13.25 < x < 14.75$
5. $x <$ \$60
6. 3:30 to 6:00
7. \$262.50
8. \$350
9. \$2 to \$4

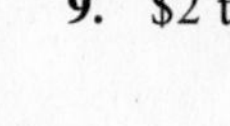

Unit 13 Posttest/pages 296-297

1.-4.

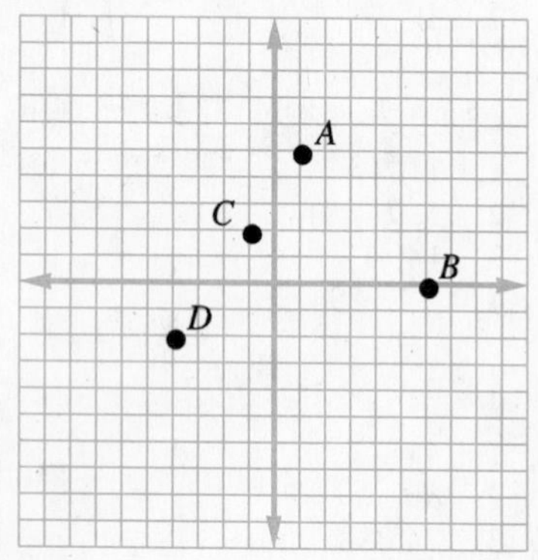

5. 7.21
6. 5
7. $-\frac{4}{9}$
8. Undefined
9. $1, \frac{1}{4}, \frac{3}{4}$
10. $-1, 3, -11$

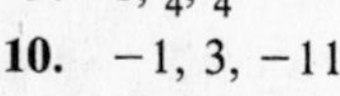

11.

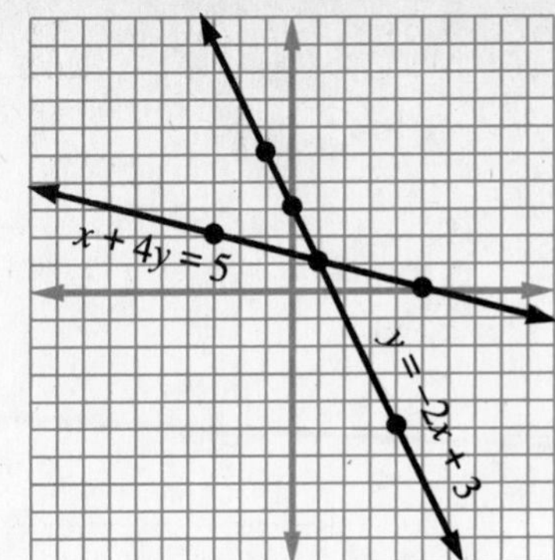

12.

4 5 6 7 8 9 10 11 12 13 14 15

13.

-4 -3 -2 -1 0 1 2 3 4 5 6 7

14.

-6 -5 -4 -3 -2 -1 0 1 2 3 4 5

15.

-8 -7 -6 -5 -4 -3 -2 -1 0 1 2 3

16. $y > \frac{-2}{5}$

$-1\ -\frac{4}{5}\ -\frac{3}{5}\ -\frac{2}{5}\ -\frac{1}{5}\ 0\ \frac{1}{5}\ \frac{2}{5}\ \frac{3}{5}\ \frac{4}{5}\ \frac{5}{5}$

17. $x < 8$

0 1 2 3 4 5 6 7 8 9 10 11

18. $y < 2$

-5 -4 -3 -2 -1 0 1 2 3 4 5 6

19. $x > 2$

-4 -3 -2 -1 0 1 2 3 4 5 6 7

20. 97

21. 8 yards